ACTA UNIVERSITATIS U
Uppsala Studies in History

46

CW00524303

JACOB ORRJE

Mechanicus

Performing an early modern persona

UPPSALA
UNIVERSITET

Dissertation presented at Uppsala University to be publicly examined in Geijersalen, Engelska parken, Thunbergsvägen 3P, Uppsala, Wednesday, 3 June 2015 at 10:15 for the degree of Doctor of Philosophy. The examination will be conducted in Swedish. Faculty examiner: Professor Thomas Kaiserfeld (Lund University, Department of Arts and Cultural Sciences).

Abstract
Orrje, J. 2015. Mechanicus. Performing an Early Modern Persona. *Uppsala Studies in History of Ideas* 46. 233 pp. Uppsala: Acta Universitatis Upsaliensis. ISBN 978-91-554-9246-5.

This thesis studies mechanics as a means of making men, rather than machines. Drawing on Swedish sources from 1700–50, it approaches mechanics as an exercise of a virtuous subject, known to his contemporaries as the "mechanicus". The *mechanicus* was a persona, consisting of expectations of the performance of mechanics that were part of the social fabric of the early modern Swedish state. The aim of this thesis is to understand how mechanical practitioners performed in relation to this persona, and how these expectations in turn were changed by actors' performances.

By studying the *mechanicus*, I take an interest in historical ways of conceiving of mechanics. Previous research on early modern mechanics has tended to relate it to modern phenomena, such as engineers, technology and industrialism, and mechanical practitioners have been considered as agents of change, who brought traditional societies into modernity. Avoiding such long narratives, this thesis presents an alternative history. By following mechanical practitioners, who staged themselves as relevant to an early modern state, I seek to understand how mechanics was presented and justified in a pre-industrial society.

The thesis is comprised of five studies. First, I discuss how mature mechanical practitioners imagined the exercise of mechanics to make a boy into a *mechanicus*. These exercises would nurture an ideal man, encompassing a range of the expected virtues of a male subject. Second, I study mechanics and geometry at the Swedish Bureau of Mines between 1700 and 1750. I show how, from having initially been associated with the building of machines and subterranean constructions, such knowledge formed the basis of a community of mathematical men of metals. Third, I analyse the letters exchanged between the mechanical practitioner Christopher Polhammar and the Swedish king Karl XII, showing how royal patronage of mechanics shaped both men. Finally, I follow the mechanical practitioner Anders Gabriel Duhre, who first succeeded and then failed to present himself as a virtuous *mechanicus* to the parliament of the Swedish constitutional monarchy of the 1720s and 1730s. Taken together, these studies show how men imagined, succeeded and failed life as a *mechanicus* in early modern Sweden.

Keywords: mechanics, performance, virtue, persona, mathematics, self fashioning, education, masculinity, age, oeconomy, cameralism, Sweden, eighteenth century, intergenerational relationships

Jacob Orrje, Department of History of Science and Ideas, Box 629, Uppsala University, SE-751 26 Uppsala, Sweden.

ISSN 1653-5197
ISBN 978-91-554-9246-5
urn:nbn:se:uu:diva-251337 (http://urn.kb.se/resolve?urn=urn:nbn:se:uu:diva-251337)

Printed in Sweden by Kph Trycksaksbolaget AB, 2015

To Lisa and Rasmus

Contents

Acknowledgements

Writing a thesis is not only about making a book, but also about creating a researcher. Beyond requiring diligent work, this process of becoming involves the accumulation of an ever-growing debt to a number of helpful friends and colleagues. I am immensely grateful to all of you!

First and foremost, I would like to express my deep gratitude to my supervisors. My main supervisor Otto Sibum, who encouraged me to explore the history of science already when supervising my Master's thesis, has over the years generously shared his knowledge in many stimulating discussions. I am likewise grateful to my co-supervisor Hjalmar Fors for revealing to me how the knowledge making of the early modern Swedish state can be surprisingly magical. A sincere thanks to both of you, for your encouragement, support and wise comments, as well as for your critical questions.

In addition, I am thankful to those of you who have read and commented on this thesis at various stages. I am especially grateful to Jens Eriksson, who has read practically every text I have written and provided me with encouraging support all these years, from his desk just next to mine. Thank you also, Andre Wakefield, for your invaluable help before, during and after my Final seminar. I am, furthermore, indebted to all of you who have provided helpful suggestions, comments, criticism and proof reading in the final stages of making this thesis: Annika Berg, Annelie Drakman, Petter Hellström, Lisa Hellman, Klara Goedecke, Måns Jansson, My Klockar Linder, Pia Levin, Anders Lundgren, Frans Lundgren, Emma Nygren, Mathias Persson, Andreas Rydberg, Göran Rydén, Kristiina Savin, David Thorsén, Linnea Tillema, Petter Tistedt, Per Widén, Sven Widmalm, and Annika Windahl Pontén. A particular thanks goes to Chris Haffenden who has edited my English here in the acknowledgements as well as on the back cover, in my introduction and in the abstract.

I wish to thank my colleagues at the Department of History of Science and Ideas as well as all the other helpful researchers at the English Park Campus in Uppsala. Thanks also to everyone at the Department of History and Philosophy of Science at Cambridge during the academic year of 2010–11, and especially Simon Schaffer, for your help and comments.

My gratitude also goes to the administrative staff at the Department in Uppsala, especially Ulla-Britt Jansson, who has gone out of her way to help me with practical matters. Likewise, thanks to the librarians at Uppsala University Library and the staff at the National Archives in Stockholm, who have brought me scores of heavy old folios.

A final thanks goes to my beloved Lisa Swedén for always being there, not only as a sharp-eyed editor but also as the person who has shown me how to write a thesis without becoming too much of a researcher.

* * *

A number of foundations have provided economic support for this thesis. My year at Cambridge was made possible by *Byzantinska stipendiestiftelsen* at Uppsala University and *Helge Ax::son Johnsons stiftelse*. Additionally, my final year of writing up this thesis has been financed through generous contributions from *Göransson-Sandvikens stipendium* at Gästrike-Hälsinge nation in Uppsala.

1. The subject of early modern mechanics

In 1727, the young mathematician Johan Mört published a mathematical textbook, translated from Latin and adapted for the youth of the Swedish state. Mört's preface explained that mathematics was not as "fruitless or breadless" as some claimed it to be. Not only was it a means of disciplining reason and imagination, it also provided "deep and respectful thoughts of GOD and his hidden wisdom". For Mört, mathematics was useful because it led young minds away from "ignorance". Mathematical education stimulated "sound and pure thoughts" and made students into "useful members of the public".[1]

To early modern Europeans, mathematics denoted a broad range of activities. The term did not only refer to "pure mathematics", such as arithmetic and geometry: it also included arts that in some way depended on numbers and figures (e.g., optics, fortification and geodesy), called "practical mathematics". This thesis is concerned with one form of practical mathematics in particular: mechanics. Most histories of early modern mechanics have discussed it as a way of building useful machines and understanding nature. Both these aspects of mechanics have been linked to long narratives of the rise of a scientific worldview, modern technology and industrialism. Mört, however, seems to imply a third, less explored, aspect of early modern mechanics, which is not so clearly part of these long narratives. He understood it as a means of crafting useful and virtuous subjects aligned to a specific social and religious order. Mört was not alone in seeing mechanics as an exercise of virtue. On the contrary, early modern authors repeatedly discussed mechanics as a useful exercise that fostered a virtuous and pious subject. To his contemporaries, this man was known as the "mechanicus".

The mechanicus is an evasive and ephemeral object of study: it consists of the expectations that came to being in the encounter between practitioners of mechanics and their contemporaries. As such, it was an ever-changing set of ideals and norms, which were both the prerequisite and result of mechanical work. The cases presented in this book draw on Swedish sources.

---

[1] "frucht- eller bröd-lös"; "diupa och wördsamma tanckar om GUD och hans fördolda Wjshet"; "okunnogheten"; "sunda och rena tanckar"; "nyttiga lemmar uti det allmänna samhäldet"; Johann Friedrich Weidler and Johan Mört (tran.): *En klar och tydelig genstig eller anledning til geometrien och trigonometrien* (Stockholm, 1727), preface. The Latin original was Johann Friedrich Weidler: *Institutiones mathematicae decem et sex pvrae mixtaeque matheseos disciplinas complexae* (Wittenberg, 1718).

In a sense, this makes it a cultural history of the mechanicus in Sweden. However, I would argue that my conclusions are broader than the geographical delimitation and that similar processes were enacted throughout early modern Europe. The mechanicus can be found in narratives of the lives and work of men who practised mechanics; it was forged through relationships between mechanically interested men; and, as we will see, it was related to the political order of the early modern Swedish state. I thus study the mechanicus from a number of perspectives, using several methodologies and categories of sources. I examine textbooks and speeches filled with imagined coming-of-age narratives of mechanics; letters in which young men presented themselves to the state administration as competent in mathematics and mechanics; correspondence where practitioners of mechanics crafted themselves in relation to the sovereign; and protocols from the Swedish parliament and rural courts that discuss matters involving mechanics and mathematics. Through these heterogeneous sources, I study the relationships between actors – in the Latin sense of "doers" – and the audiences who interpreted their actions. I thus take an interest not only in what has been written about mechanics, but also for whom it was written and for what reasons.

I examine the processes by which men formed themselves in relation to the audiences' and actors' expectations that constituted the mechanicus, and how these expectations were changed by performances of mechanics. These "mechanical practitioners" (i.e., actors who were, or were trying to be, recognised as knowledgeable in this field) should not be seen as agents of change in opposition to an early modern order, but as men working to meet the expectations of their contemporaries. Therefore, the mechanicus should be related to the social and natural orders of his time, as well as to religion, politics and epistemologies (i.e., frameworks of making and categorising knowledge). I thus examine how mechanical exercises shaped and made early modern men, asking questions such as: How was the mechanicus a part of the early modern Swedish state? How did men present themselves as trustworthy and valuable subjects through mathematics and crafts? How did they in turn shape the expectations of who a mechanicus was? How can we understand mechanical practitioners' successes, and failures, to meet the expectations of contemporary audiences?

Methodological considerations and delimitations

This thesis shifts the perspective on early modern mechanics in two ways. The first shift is a *methodological* one, from long narratives to a synchronic history. The study examines historical actors' expectations of mechanics during the first half of the eighteenth century. As such, it diverges from studies that analyse early modern mechanics in order to construct long narratives of the making of modern phenomena such as industrialism, professional engineers or modern technocratic states. In order to study eighteenth-century mechanics as aligned with its time, I make a methodological pledge of abstinence from such long narratives and from stabilising actors' categories by using analytical terms (i.e., modern terms that were alien to the historical actors themselves). The second shift is a *geographical* one, from England, which has often been staged as the source of modern industrialism, to the poor Northern-European state of Sweden. Chronologically, my cases range from the beginning of the 1700s to the mid century, a period during which Sweden underwent drastic governmental changes. From having been an absolutist state, in the 1720s it became a constitutional monarchy. The Swedish case thus highlights how expectations of mechanical practitioners were tied to shifting early modern political orders. Furthermore, these methodological and geographical shifts are interrelated: a study of the mechanicus in the Swedish state facilitates telling histories of early modern mechanics other than those of industrialisation or the rise of modern science.

Over the twentieth century, the British industrial revolution became enrolled in liberal ideology and historiography, which described early modern mechanical practitioners as agents of change. They thus came to be interpreted as "entrepreneurial individuals operating in the private sphere, as opposed to the sphere of the state, who generated rational and useful knowledge."[2] The industrial revolution, and with it the interpretation of the historical role of mechanical practitioners, became "embedded within debates over competing systems of political economy, primarily liberal democracy (free trade) versus socialism (state regulation)."[3] This liberal narrative still shapes influential scholars' interpretations of the relationship between eighteenth-century knowledge and the industrial revolution. For example, Joel Mokyr has argued for a position that attributes Western industrial development to both individual inventors and culture. For him, "technological

---

[2] William J. Ashworth: "The British industrial revolution and the ideological revolution. Science, neoliberalism and history", *History of science* 52:2 (2014), 182; see also William J. Ashworth: "The ghost of Rostow. Science, culture and the British industrial revolution", *History of science* 46:3 (2008), 249–74. For a discussion of the role of such narratives in recent history of science, see Andre Wakefield: "Butterfield's nightmare. The history of science as Disney history", *History and technology* 30:3 (2014), 232–51. Wakefield provided me with the manuscript of this article, which helped me sharpen the methodological arguments of this thesis.

[3] Ashworth: "The British industrial revolution and the ideological revolution", 178. Compare also to nineteenth-century heroic conceptions of invention, discussed in Christine MacLeod: "Concepts of invention and the patent controversy in Victorian Britain", in Robert Fox (ed): *Technological change. Methods and themes in the history of technology* (Amsterdam, 1996), 150–3. See also Christine MacLeod and Alessandro Nuvolari: *"The ingenious crowd". A critical prosopography of British inventors, 1650–1850* [working paper] (Eindhoven, 2005); Christine MacLeod: *Heroes of invention. Technology, liberalism and British identity, 1750–1914* (Cambridge, 2007); David Edgerton: *The shock of the old. Technology and global history since 1900* (New York, 2007), x.

change involves an attack by an individual on a constraint that everyone else takes as given."[4] Mokyr sees the historical application to rational production of useful mechanical knowledge, found in small subcultures of knowledgeable men, as a key explanation of modern industrialism. In his view, small groups of mechanical practitioners were the cornerstones of an industrial enlightenment. In his account, this small subculture of rational men brought about the affluence of the post nineteenth-century Western world.[5] Similarly, Margaret C. Jacob has recently argued that "knowledge of the physical universe gave entrepreneurs a singular advantage". By participating in contemporary "scientific culture", such actors could approach mechanical production in ways previously deemed "irrelevant to *homo economicus* as classically formulated."[6]

Mokyr and Jacob both see mechanical practitioners as acting unexpectedly in relation to contemporary culture. Here, I present an alternative narrative, where the self-presentations of mechanical practitioners are seen neither as opposing, nor as transgressing, contemporary audiences or cultures. Instead, I study these presentations as made in relation to the shifting and complex expectations of early modern audiences. By taking this approach, I highlight the heterogeneous roles of early modern mechanics, and especially such roles that diverge from the assumption that mechanics should be studied as a new form of invention that emanated from England. In his recent *The limits of matter* (2015), Hjalmar Fors has criticised such unidirectional models, which proceed from an often "unconscious, centrist historiographical assumption: that eighteenth-century invention proceeded from England." Instead, his study balances "the substantial literature on the mechanical and technical aspects of early modern mining."[7] This study draws on Fors' critique of centralist assumptions. However, I also argue that early modern mechanics does not have to be studied as a form of "technology" or "innovation". Instead, an interest in multiple and diverging meanings of mechanics motivates this study of the mechanicus in the early modern Swedish state. There, a mechanicus was expected to be someone very different from the entrepreneurial individual, or innovator, who Mokyr and Jacob idolise as the maker of a modern order of production.

---

[4] Joel Mokyr: *The lever of riches. Technological creativity and economic progress* (New York, 1990), 9.

[5] Joel Mokyr: *The gifts of Athena. Historical origins of the knowledge economy* (Princeton NJ, 2005), 52–3.

[6] Margaret C. Jacob: *The first knowledge economy. Human capital and the European economy, 1750–1850* (Cambridge, 2014), 221. Andre Wakefield has forcefully, and convincingly, criticised Jacob and Mokyr, showing how their histories draw on "Cold War modernization theory", in Wakefield: "Butterfield's nightmare", 239–44.

[7] Hjalmar Fors: *The limits of matter. Chemistry, mining, and Enlightenment* (Chicago, 2015), 153, see also 9. During my work I have had access to Fors' study in manuscript form, which has been especially helpful in framing Chapter 3.

Early modern mechanics as performance

Thus far, I have established that the focus of this thesis is an object of study called the "mechanicus". I have also made a pledge of abstinence, not from all methodology but from approaches that understand the mechanicus in relation to long narratives. The alternative to such an approach is hardly, as it were, a method of no method. In order to understand the expectations of a mechanicus, I use a theoretical approach with which it is possible to see the performances by which historical actors attributed him meaning.

The mechanicus was a *persona*. As pointed out by H. Otto Sibum and Lorraine Daston, a persona can be understood as a "social species": a way of being and knowing, found in between the individual and the collective. As such, personas are "not individuals, nor are they simply stereotypes or social roles."[8] A persona is a mask, but not in the modern sense of something that distances an individual from his or her essential identity. Instead, it is a mask that genuinely transforms the individual who puts it on. Consequently, personas created "possibilities of being in the human world, schooling the mind, body and soul in distinctive and indelible ways."[9] Like other personas, the mechanicus was not a role that a mechanical practitioner could simply step into, while leaving their "true self" intact: it transcended the individual. I see the mechanicus as a set of expectations that provided a framework of meaning to mechanical practitioners – a framework that enabled and disabled them from taking certain actions. It was an ever-changing assemblage of demands and expectations, which gave meaning to actors' performances as well as audiences' reactions, and which transformed the mechanical practitioners who performed in relation to it.

Residing in between the individual and the collective, the expectations that constituted the mechanicus inhabited both a *structural* and a *relational* level. On the one hand, on a structural level these expectations were part of the fabric of the social and political order of the Swedish state and of early modern categories of knowledge making. As such, the mechanicus was an expression of an early modern "social epistemology": it was a means of

---

[8] Lorraine Daston and Otto Sibum: "Introduction. Scientific personae and their histories", *Science in context* 16:1 (2003), 3.

[9] Quotation from ibid., 4. On the concept of persona in an early modern context, see also Conal Condren, Stephen Gaukroger and Ian Hunter: "Introduction", in Conal Condren, Stephen Gaukroger, and Ian Hunter (eds): *The philosopher in early modern Europe. The nature of a contested identity* (Cambridge, 2006), 1–16; Stephen Gaukroger: "The persona of the natural philosopher", in Condren, Gaukroger, and Hunter (eds): *The philosopher in early modern Europe*, 17–34. On the holding of an early modern office as a persona, see Conal Condren: "The persona of the philosopher and the rhetorics of office in early modern Europe", in Condren, Gaukroger, and Hunter (eds): *The philosopher in early modern Europe*, 66–89. Sorona Corneanu discusses a persona as "an exemplary identity wrought by intellectual, moral, and even corporeal disciplines, one that represented an *office* (sometimes a noninstitutionalized one) in specific cultural spaces"; Sorona Corneanu: *Regimens of the mind. Boyle, Locke, and the early modern cultura animi tradition* (Chicago, 2011), 7.

categorising various forms of knowledge making, which was inseparable from questions of political power, identity or relationships between social groups.[10] On the other hand, the mechanicus was enacted, shaped and reinterpreted in actors' *relational performances*. By studying such performances, it is possible to see mechanical practitioners as being shaped in relation to political power as well as to contemporary audiences. The performances that I have studied were relational in two senses. First, they were always made in relation to an audience. Second, they were enacted in long-term transformative relationships. As such, mechanics was a means of weaving together and transforming early modern men through relations of difference and similarity. These differences and similarities were manifested through a number of social distinctions such as gender, age and social status.

Although I see these relationships as permeated by, and as expressions of, contemporary social epistemology, I do not argue that mechanical practitioners were following any form of static script. There was always interpretative flexibility in their performances and there always existed means for them to relate tactically, in Michel de Certeau's meaning of the term, to contemporary expectations. Thus, mechanical practitioners not only adapted to the mechanicus: they also appropriated, stretched and reinvented this persona. These tactical performances were made up of "heterogeneous elements", pieced together in a seizing of the moment.[11] The elements of mechanical performances consisted of a number of *techniques*. In his *Les techniques du corps* (1934), Marcel Mauss defines techniques as "the ways in which men [and women], society by society, in a traditional fashion, know how to use their bodies." Mauss points out that techniques are part of the social world of the communities that we live in. The acts of an individual are formed "by all his education, by all of the society which he is part of, in the space that he occupies." Mauss thus argues that as much authority and social tradition are involved when transmitting bodily actions as when mediating language.[12] Therefore, human beings constantly create, break and reaffirm

---

[10] Ken Alder: *Engineering the revolution. Arms and enlightenment in France 1763–1815* (Princeton NJ, 1997), 56. With this term I do not denote the field of philosophy discussing principal questions of the relationship between knowledge and "the social". See for example Frederick F. Schmitt: *Socializing epistemology. The social dimensions of knowledge* (Lanham MD, 1994).

[11] Michel de Certeau defines the actor performing tactically as someone who is "always on the watch for opportunities that must be seized 'on the wing.' Whatever it wins, it does not keep."; Michel de Certeau: *The practice of everyday life* (Berkeley & Los Angeles, 1984), xix.

[12] "les façons dont les hommes, société par société, d'une façon traditionnelle, savent se servir de leur corps."; Marcel Mauss: *Les techniques du corps* (Chicoutimi, 2002 [1934]) <http://classiques.uqac.ca/classiques/mauss_marcel/socio_et_anthropo/6_Techniques_corps/techniques_corps.pdf> [accessed 27 April 2012], 5. "par toute son éducation, par toute la société dont il fait partie, à la place qu'il y occupe."; ibid., 10.

Similarly, Liliane Hilaire-Pérez has argued that "skills become embedded within specific communities because of their needs and constraints, their habits and symbols, and their territories." She points out that "'techniques' are answers to specific needs and expectations. Their applications are not universal; they belong to a world of diversity, contingency, and heterogenei-

social affiliations through the techniques by which they engage with and exist in the world.

I approach a wide range of techniques carried out by mechanical practitioners: for example, socialising, writing, talking, crafting, drawing and calculating. All these techniques were enabling but also disabling, they opened up new possibilities for engaging with social groups and the world, while they also restricted other ways of acting. Mechanical practitioners obviously engaged with techniques considered directly related to the making of mechanical knowledge, especially crafts and mathematics (interpreted in the broad early modern sense). However, they also routinely performed a wider range of techniques, such as writing and drawing, by which they reflected upon mechanics and presented it to contemporary audiences. I do not see these two categories of techniques as essentially different from each other. Instead, I see all of them as means of performing mechanics. The techniques constituted what Sibum has termed a "gestural knowledge". In other words, they constituted a form of knowledge that was "united with the actor's performance of work" and changed "according to the specific kinds of performance".[13]

But then what do I mean by *performances* of mechanics? A starting point for this thesis is the simple observation that, when historical actors engaged with mechanics, they also made themselves in the eyes of their contemporaries. Mechanical techniques were thus not only a means of making knowledge and building machines, but also, to borrow a term from Judith Butler, *identificatory practices*. That is, the mechanicus was "in no way a stable identity or locus of agency from which various acts proceed; rather, it is an identity tenuously constituted in time – an identity instituted through a *stylized repetition of acts*." Mechanical performances were transformative, and the mechanicus was "instituted through the stylization of the body and, hence, must be understood as [involving] bodily gestures, movements, and enactment of various kinds".[14] Thus, I see performances of mechanics as means

ty." Also, the spaces through which these techniques were circulated "were not neutral or passive; instead, they always adapted and translated the techniques they conveyed or received." Liliane Hilaire-Pérez: "Dissemination of technical knowledge in the Middle Ages and the early modern era. New approaches and methodological issues", *Technology and culture* 47:3 (2006), 537.

[13] H. Otto Sibum: "Reworking the mechanical value of heat. Instruments of precision and gestures of accuracy in early Victorian England", *Studies in history and philosophy of science. Part A* 26:1 (1995), 76 note 8; H. Otto Sibum: "Les gestes de la mesure, Joule, les pratiques de la brasserie et la science", *Annales HSS* July–October, 1998; H. Otto Sibum: "Experiencing experiment. Gestural knowledge and scientific change in early 19th-century Victorian culture", *Thought & culture* 10 (2011), 38–55. See also Simon Schaffer: "Experimenters' techniques, dyers' hands, and the electric planetarium", *Isis* 88:3 (1997), 458.

[14] Judith Butler: "Performative acts and gender constitution. An essay in phenomenology and feminist theory", in Henry Bial (ed): *The performance studies reader* 2nd edition (London & New York, 2007), 187. On presentation and performance of social attributes, see also Erving Goffman: *The presentation of self in everyday life* (London, 1990), 81, and passim. How knowledge making is related to the making of the scientist or the artisan through repetitious bodily acts has

of knowing, acting and living in the world. These performances placed moral demands on the audience, which obliged them to treat the actor in the manner that a mechanicus had a right to expect.[15] Again, these demands are meaningful both as part of an early modern social epistemology and in relation to a particular audience and a specific situation.

By studying early modern mechanics as performances, made for contemporary audiences, I delineate the historical processes that ascribed meaning to mechanical practitioners. The question of who the mechanicus was is thus not something that can be defined in this introduction: it is a question that must be answered in relation to the specific institutions and relationships through which the mechanicus was made, remade and unmade.

Actors' categories

The term "mechanicus", or "mechanici" in the plural, was routinely used in the seventeenth and eighteenth centuries to signify men who were deemed skilled in mechanics. Certain early modern Swedes were also identified as an *ingenieur*.[16] However, the term *ingenieur* seems to have been used less often than the term "mechanicus". Generally, whereas a mechanicus was a man defined by his able performance of mechanics, an *ingenieur* was someone with a specific position in the state. It might seem convenient to equate the term *ingenieur* with the modern "engineer". However, when we approach techniques of the past, our modern understanding of terms such as "engineers", "invention" or "technology" can be problematic. To study a term such as the "mechanicus", which is less easy to associate with modern categories, and which historical actors themselves understood as defined by men's performances, is a way to avoid these problems. This history of the

been discussed by Sibum: "Les gestes de la mesure, Joule, les pratiques de la brasserie et la science"; Pamela H. Smith: *The body of the artisan. Art and experience in the scientific revolution* (Chicago, 2004), 18–20. The topic has also been treated by the texts in Christopher Lawrence and Steven Shapin (eds): *Science incarnate. Historical embodiments of natural knowledge* (Chicago, 1998); See especially: Christopher Lawrence and Steven Shapin: "Introduction. The body of knowledge", in Lawrence and Shapin (eds): *Science incarnate*, 1–19; Andrew Warwick: "Exercising the student body. Mathematics and athleticism in Victorian Cambridge", in Lawrence and Shapin (eds): *Science incarnate*, 288–323.

[15] For a discussion of the moral demands of performances, see Goffman: *The presentation of self in everyday life*, 24.

[16] For example, in 1692 the Captain engineer [*Capitein Ingenieur*] in the fortification Otto Barthold Scholl published a treatise on geometry in Swedish; Barthold Otto Schmoll: *Kort anledning till geometrien, hwar efter officerarne wid Hans Kongl. May:tz infanterie här i Swerige till mathesin och fortification* (Stockholm, 1692). Also, in the early eighteenth century Petrus Tillæus was the city engineer [*Stads Ingenieur*] of Stockholm as seen on his detailed map over Stockholm: Petrus Tillaeus: "General charta öfwer Stockholm med malmarne åhr 1733" (Stockholm, 1733). Svante Lindqvist has described the "mechanicus" as a "universal title", in relation to the term "ingenieur" which in the early eighteenth century "was used principally as a title of officers in the Fortification Corps"; Svante Lindqvist: *Technology on trial. The introduction of steam power technology into Sweden, 1715–1736* (Uppsala, 1984), 15.

mechanicus is thus not a history of the rise of a profession, or of the modern engineer. It is the history of men who made themselves in relation to contemporary expectations.

A number of historians of technology have argued for the importance of being sensitive to historical actors' own categories. For example, they have pointed out the problems in defining engineers or mechanical practitioners as inventors, because this equates mechanical work with the act of invention.[17] Scholars of early modern mechanics have highlighted similar issues. For example, Jim Bennett has pointed out how the historical understanding of early modern mechanics has been crippled by twentieth-century historians' tendency to reduce the complexity of this field to the two modern categories of science and technology. The introduction of these analytical concepts has made historians see a divide between a mechanistic natural philosophy and mechanical and mathematical practitioners, which was not obvious to historical actors. As a consequence, historians have established artisans and scholars as two categories a priori and have identified historical actors, or sometimes specific parts of an actor's work, as belonging to one of the two.[18] Such studies, which analytically categorise historical phenomena as either science or technology, tend to generate questions of influences and consequences ripe with teleology: did modern science originate in the world of craftsmen, or was modern technology the result of a rationalisation and mathematisation of artisanal knowledge? In historical studies of periods prior to the use of these modern concepts, science and technology tend to encourage questions about origins, and thus to make historians blind to processes other than those that gave rise to these opposing terms themselves. Sometimes, such use even compels historians to write histories in which these terms are static and a-historical, and in which the performances of historical actors are interpreted as made in relation to these categories, as they are understood today. Either way, by using such distinctions, we end up with questions as to how early modern mechanics made the industrial world. These are questions that early eighteenth-century actors themselves

---

[17] This point has recently been made in David Edgerton: "Innovation, technology, or history. What is the historiography of technology about?", *Technology and culture* 51:3 (2010), 683. See also Eric Schatzberg: "Technik comes to America. Changing meanings of technology before 1930", *Technology and culture* 47:3 (2006), 486, 512; Ruth Oldenziel: *Making technology masculine. Men, women, and modern machines in America, 1870–1945* (Amsterdam, 1999), 14, 46. Christine MacLeod has made similar arguments about the interpretation of historical patent systems. See Christine MacLeod: *Inventing the industrial revolution. The English patent system, 1660–1800* (Cambridge, 1988), 5, 158–81. For a critical discussion of the view of engineers as inventors, see Edgerton: *The shock of the old*, xv. Also, Andre Wakefield makes a similar call for what he calls a history "that looks forward instead of back"; Andre Wakefield: *The disordered police state. German cameralism as science and practice* (Chicago, 2009), 23; Andre Wakefield: "Leibniz and the wind machines", *Osiris* 25:1 (2010), 172–3, 188.
[18] Jim Bennett: "The mechanics' philosophy and the mechanical philosophy", *History of science* 24 (1986), 6.

obviously knew very little about, and which can hardly be said to have shaped their own understanding of their actions.

A similar point is made by the authors in *The mindful hand* (2007), a collection of studies with a common aim of understanding material and knowledge production as "a single, hybrid affair in which the work of the head and of the hand formed a complex whole." These are not studies of "the relationship between science and technology, or of the relationship between theory and application."[19] Moreover, they refute essentialist distinctions of theory's superiority over practice, because it is to conjure forth an opposition between "on one hand the idealising method whereby philosophers mechanised the world-picture and, on the other, an applied set of practices whereby mechanics improved their techniques".[20] Instead, they treat early modern mathematics, including mechanics, as an amalgam of what we today might call science and technology.[21] In their view, the "examination of nature was just as likely to involve and contribute to material production as invention was to the production of knowledge."[22] Similarly, a number of recent studies have approached mechanical practitioners as go-betweens – that is, as actors who translated and mediated between contemporary cultures of craftsmen and natural philosophers.[23]

My thesis takes many cues from all these studies that question how historians of the nineteenth and twentieth centuries have conceptualised historical mechanical work. It is important to recognise that boundaries that today might appear settled – such as those between science and technology, or theory and practice – may have been drawn differently, been more fluid, or even been non-existent to historical actors. Such categories, and the relationships between them, should not be seen as preceding actors' performances, but as products of them. Therefore, this thesis sets out neither to reduce the field of mechanics to modern categories such as science and technology, nor to show that mechanics was an amalgam of what today are categorised by these terms. Instead, the boundaries between, for example, theory and practice (to the actors known as *theoria* and *praxis*) were constant-

---

[19] Lissa Roberts, Simon Schaffer, and Peter Dear (eds): *The mindful hand. Inquiry and invention from the late Renaissance to early industrialisation* (Amsterdam & Bristol, 2007), x.

[20] Lissa Roberts: "Introduction. Workshops of the hand and mind", in Roberts, Schaffer, and Dear (eds): *The mindful hand*, 6.

[21] Roberts, Schaffer and Dear (eds): *The mindful hand*, xix.

[22] Ibid., x.

[23] Lissa Roberts: "Full steam ahead. Entrepreneurial engineers as go-betweens during the late eighteenth century", in Simon Schaffer et al. (eds): *The brokered world. Go-betweens and global intelligence, 1770–1820* (Sagamore Beach MA, 2009), 193–238; Lissa Roberts: "Geographies of steam. Mapping the entrepreneurial activities of steam engineers in France during the second half of the eighteenth century", *History and technology* 27:4 (2011), 420; Chandra Mukerji: *Territorial ambitions and the gardens of Versailles* (Cambridge, 1997); Chandra Mukerji: *Impossible engineering. Technology and territoriality on the Canal du Midi* (Princeton NJ, 2009).

ly redrawn by actors who struggled to perform in relation to the expectations of contemporary audiences.

From long narratives to synchronic history

By the late twentieth century, historians of science and technology abandoned writing simple chronicles and linear determinist models of the relationship between knowledge, technology and affluence. Instead, in the wake of the new sociology of science of the 1970s – interested in scientific practice as a part of society – the focus of historians of science has shifted to science as a social and collective process. This shift has fundamentally enriched our understanding of science as constituted by social, material and historical practices.[24] Working in this mode of historical inquiry, an increasing number of studies from the 1980s onwards have re-examined the role of early modern mechanics, machines and mechanical practitioners. They have shown how such actors were part of a history of values, products, objects, techniques and procedures, which together gave legitimacy to new objects according to contemporary expectations.[25]

Among these works on eighteenth-century mechanics, two stand out as especially relevant to my study: Ken Alder's *Engineering the revolution* (1997) and Svante Lindqvist's *Technology on trial* (1984). Alder studies an "engineering technological life" in France during the *ancien régime* and under the

---

[24] Notable examples of this shift are Steven Shapin and Simon Schaffer: *Leviathan and the air-pump. Hobbes, Boyle, and the experimental life* (Princeton NJ, 1985); Bruno Latour: *The pasteurization of France* (Cambridge MA, 1988). These works were focusing on issues similar to those of works in the sociology of science from the 1970s, promoted by on the one hand David Bloor and on the other hand Harry Collins, see for example David Bloor: "Wittgenstein and Mannheim on the sociology of mathematics", *Studies in history and philosophy of science. Part A* 4:2 (1973), 173–91; H. M. Collins: "The seven sexes. A study in the sociology of a phenomenon, or the replication of experiments in physics", *Sociology* 9:2 (1975), 205–24.
   In the 1980s, a number of sociologists of science proposed the usefulness of not establishing differences between science and technology *a priori*. See for example Wiebe Bijker and Trevor Pinch: "The social construction of facts and artifacts. Or how the sociology of science and the sociology of technology might benefit each other", in Wiebe Bijker, Thomas Hughes, and Trevor Pinch (eds): *The social construction of technological systems. New directions in the sociology and history of technology* (Cambridge MA, 1987), 17–50; Bruno Latour: *Science in action. How to follow scientists and engineers through society* (Cambridge MA, 1987).
[25] Many studies analyse the historical and social specificities of early modern engineers and inventors, both in a synchronic and diachronic context. Noteworthy examples are: Lindqvist: *Technology on trial*; MacLeod: *Inventing the industrial revolution*; Larry R. Stewart: *Rise of public science. Rhetoric, technology and natural philosophy in Newtonian Britain, 1660–1750* (Cambridge, 1992); Antoine Picon: *French architects and engineers in the Age of Enlightenment* (New York, 1992); Hélène Vérin: *La gloire des ingénieurs. L'intelligence technique du XVIe au XVIIIe siècle* (Paris, 1993); Alder: *Engineering the revolution*; Liliane Hilaire-Pérez: *L'invention technique au siècle des Lumières* (Paris, 2000), 25, 36–9; Margaret C. Jacob and Larry Stewart: *Practical matter. Newton's science in the service of industry and empire, 1687–1851* (Cambridge MA, 2004); Jean-François Gauvin: *Habits of knowledge. Artisans, savants and mechanical devices in seventeenth-century French natural philosophy* (Cambridge MA, 2008); MacLeod: *Heroes of invention*.

government of the French revolution. This technological life, "a coherent and ideological world which gives purpose and meaning to a set of material objects", was a French absolutist context that was in some ways similar to the Swedish one studied here. Alder shows how, in eighteenth-century France, mechanics was not an activity of entrepreneurial individuals, but instead one of civil servants submitted to the state.[26] As such, his study highlights the different roles of mechanical practitioners in various early modern European states, and emphasises that we should not see France or Sweden through the models constructed by researchers who have, predominantly, studied British sources.

Svante Lindqvist's *Technology on trial* is an early and important Swedish study, which shows the importance of being sensitive to the social and material organisation of eighteenth-century science. As part of the developments in the history of science from the 1980s, it adjusts the traditional narrative of mechanics in the eighteenth-century Swedish state. Lindqvist argues that the (failed) introduction of English "steam engine technology" can only be understood by examining social and material differences between England and Sweden. *Technology on trial* is the most in-depth study of Swedish eighteenth-century mechanics to date, and provides important historical understanding of the work of mechanical practitioners. It shows how the introduction of modern mechanical production was neither a linear process nor brought about by exceptional individuals. Lindqvist writes an eighteenth-century history of technology that disentangles the study of technology from hagiographies of genius inventors, and which is highly sensitive to the material and social aspects of eighteenth-century institutions.[27]

---

[26] Alder: *Engineering the revolution*, xii.

[27] Lindqvist: *Technology on trial*. In addition to Lindqvist, Mikael Hård connects Polhem's theories to contemporary philosophies, such as Cartesianism, which he in turn links to Polhem's mechanical practice, Mikael Hård: "Mechanica och mathesis. Några tankar kring Christopher Polhems fysikaliska och vetenskapsteoretiska föreställningar", *Lychnos* (1986), 55–69. Per Dahl's work on the Swedish seventeenth-century "technological education" of Olof Rudbeck and his technical enterprises is another study taking historical specificity seriously, Per Dahl: *Svensk ingenjörskonst under stormaktstiden. Olof Rudbecks tekniska undervisning och praktiska verksamhet* (Uppsala, 1995). Also, David Dunér has recently analysed the thinking of Polhem and Swedenborg in ways inspired by cognitive science, and he sees the intellectual work of these mechanical practitioners as formed by their time, see David Dunér: *Världsmaskinen. Emanuel Swedenborgs naturfilosofi* (Lund, 2004); this book is also published in English as David Dunér: *The natural philosophy of Emanuel Swedenborg. A study in the conceptual metaphors of the mechanistic world-view* (Dordrecht, 2013); David Dunér: *Tankemaskinen. Polhems huvudvärk och andra studier i tänkandets historia* (Nora, 2012).

The use of eighteenth-century inventors to form a Swedish national narrative has been discussed by Ingmarie Danielsson Malmros: *Det var en gång ett land. Berättelsen om svenskhet i historieläroböcker och elevers föreställningsvärldar* (Lund, 2012), 140–4. Such a narrative can perhaps most clearly be seen in Samuel E. Bring: "Bidrag till Christopher Polhems lefnadsteckning", *Christopher Polhem. Minnesskrift utgifven av Svenska teknologföreningen* (Stockholm, 1911); translated into English as Samuel E. Bring; William A. Johnson (tran.): *Christopher Polhem. The father of Swedish technology* (Hartford CT, 1963). In contrast to recent scholarly work on Swedish eighteenth-century me-

These earlier works provide important starting points for my study of the mechanicus in the early modern Swedish state. Still, they also exhibit a methodological ambivalence, typical of late twentieth-century studies.[28] Ian Hacking has discussed the double nature of such studies, in a review of the second edition of Simon Schaffer and Steven Shapin's *Leviathan and the air-pump* (1985). There, Hacking argues that despite "its vigorous profession of anti-anachronism," the book is "thoroughly whig history." The term "whiggism" should not, as it usually is, be seen as pejorative. As Hacking points out, perhaps these studies need to be whiggish, "because there is no way of unthinking the experimental style of reasoning that came into being."[29] This tension between anti-anachronism and whiggism is illustrative of the history of science and technology from the late twentieth century. Generally, the historically sensitive studies from this time are motivated not primarily by a wish to understand the meaning that actors ascribed to early modern knowledge, but by an interest in discovering the origins of modern phenomena. Consequently, they examine the rise of scientific objectivity and public science, modern ways of disciplining perception and the establishment of the organisational structures of modern science.

Similarly, while Alder and Lindqvist are highly sensitive to historical contexts and phenomena, their ultimate aim is to understand processes such as industrialisation and the rise of technology. Alder writes a history that places eighteenth-century engineers in a long narrative, which is made possible by a number of analytical interventions. He sees his historical actors as representatives of an ahistoric category of men: as "a subspecies of *Homo faber*, engineers are engaged in one of the most quintessential human activities." Engineers are then related to the process of industrialisation, or, as Alder puts it, "engineers have come to play a dominant role in the pattern of Western industrialization."[30] By placing eighteenth-century engineers in this long narrative, Alder aims to show how "modern technological life is a political creation, and [how] we can read in its artifacts the history of the struggles and negotiations which gave it birth."[31] Similarly, Lindqvist ultimately aims to understand the conditions that give birth to specific

---

chanics, more popular works have not fundamentally challenged this historiography. See for example: Michael Lindgren and Per Sörbom: *Christopher Polhem 1661–1751. "The Swedish Daedalus"* (Stockholm, 1985); Michael Lindgren: *Christopher Polhems testamente. Berättelsen om ingenjören, entreprenören och pedagogen som ville förändra Sverige* (Stockholm, 2011).

[28] On whiggism and late twentieth-century history of science, see Wakefield: "Butterfield's nightmare", 237–9.

[29] Ian Hacking: "Artificial phenomena. Essay review of *Leviathan and the air pump*", *The British journal for the history of science* 24:02 (1991), 238. For a more thorough discussion of whiggism in the history of science, as well as a discussion of what separates whiggish and triumphalist narratives, see Hasok Chang: "We have never been whiggish (about phlogiston)", *Centaurus* 51:4 (2009), 239–64.

[30] Alder: *Engineering the revolution*, 11.

[31] Alder: *Engineering the revolution*, xii.

technology. Using "the huge bronze cylinder of the Dannemora engine" he aims to find "an image of early eighteenth-century Sweden."[32]

Also, Lindqvist aims to understand the "transfer of technology" to new places. He follows what he sees as a unidirectional technology transfer from England to Sweden, in order to understand why technology could, or could not, move from there to a new place of adoption.[33] The image reflected in the bronze cylinder thus, in many respects, becomes a comparison with England, and ultimately with modernity. In *Technology on trial*, the steam engine becomes more than a material artifact: implicitly, it is also a metaphor for a modern industrialism, based on fossil fuels. Consequently, Lindqvist's study aims to understand why eighteenth-century England was industrialised, and why Sweden, at the same time, was not.

Shapin has defended the wish to understand the present as a legitimate aim of historical studies. Such accounts can show what "we think led on – never directly or simply, to be sure – to certain features of the present in which, for certain purposes, we happen to be interested". On a principle level, I agree that there "is nothing at all wrong about telling such stories".[34] Historians *must* choose, and justify, how and why they interact with, and interpret, historical phenomena. Such choices are never analytically innocent and they have consequences, of which one must be aware: Which histories do they have us tell, and which do they make us blind to? What are the consequences of aiming to do two things at once: of wishing to understand both the historical specificity of historical phenomena *and* the role of these phenomena in long narratives? While it is obvious that such long narratives can inform us of processes forming the present, a double-headed methodology requires historians not only to study actors' performances in relation to contemporary expectations, but also in relation to the expectations of posterity.

Whereas I share the interest in historical specificity of Alder and Lindqvist I believe that it is possible to say more about early modern understandings of mechanical practitioners by avoiding the long narratives, of which the historical actors were unaware. Although it is certainly possible to write a history of mechanics that either takes the long-term effects of mechanics or actors' own understanding into account, one cannot fully commit to both these aims at once. To study the mechanicus, as it was performed and understood by historical actors, is a conscious methodological choice, which enables me to focus on the issues at hand while ignoring others. In other words, I study mechanical performances in relation to the expectations of contemporary audiences, and not in relation to posterity. By making this choice, I consciously turn a blind eye to the role of early modern me-

[32] Lindqvist: *Technology on trial*, 13–14.

[33] Ibid., 11–12.

[34] Steven Shapin: *The scientific revolution* (Chicago, 1996), 7.

chanics in long narratives, in order to see actors' own understanding of events in a clearer light. Therefore, this study will say less about long-term historical change, and more about the meaning given to the mechanicus at a specific time and place. Daniel Roche has identified the wish to understand "how change became possible in a world that saw itself as stable, change-less" as a central problem in eighteenth-century studies.[35] In this thesis, however, I am more interested in how mechanical practice, today a part of several narratives of historical change, was justified by historical actors because it carried the promise of maintaining, or ever perfecting, the status quo.

A social epistemology of early modern mechanics

Then how did early moderns understand mechanics and the mechanicus? In early modern Europe mechanics was not only considered an art of machine building, but also "a means of understanding larger social forces, or abstract entities at play in the world."[36] Machines were not only a means of material affluence, but also provided a coherent and orderly structure by which the order of nature as well as a social order could be understood in relation to human artifice. This structure was entrenched in specific language communities, and formed networks of metaphors that asserted "meaningful similarities" between machines and a range of other phenomena. When studying these networks, it is important to perceive "the flexibility of all terms in a metaphorical identification."[37] An important starting point for this thesis is the observation that the mechanicus, like the machinery he built, was a meaningful part of various political orders. Therefore, mechanical performances could inspire early modern audiences to dream of an affluent and orderly future, and some sovereigns even loosened their purse strings for those who managed to meet their expectations of useful mechanical work.[38]

The political meanings of mechanics, and consequently of a mechanicus, were not uniform in early modern Europe. Instead, these political connotations varied along with the structure of the states in which mechanics was performed. In the absolutist states of the European continent, authors put an emphasis on the hierarchical structure of mechanical apparatus. There,

---

[35] Daniel Roche: *France in the Enlightenment* (Cambridge MA, 1998), 6.
[36] Jonathan Sawday: *Engines of the imagination. Renaissance culture and the rise of the machine* (London, 2007), 54.
[37] Peter Dear: *Discipline & experience. The mathematical way in the scientific revolution* (Chicago, 1995), 151–3, quotes from page 151 and 152. In his use of a network model of metaphor, Dear follows the model proposed by Barry Barnes: "On the conventional character of knowledge and cognition", *Philosophy of the social sciences* 11:3 (1981), 303–33.
[38] Jonathan Sawday: *Engines of the imagination*, 101. See also Wakefield: "Leibniz and the wind machines".

clock-metaphors became a means of imagining a rationally ordered society, in which each subject performed faithfully and where power emanated from a single sovereign. In the constitutional state of early eighteenth-century Britain, on the other hand, self-regulating mechanisms legitimated a political system of checks and balances. There, the point that no machinery could function without supervision and repairs became an argument against an autocratic social order.[39] There was thus interpretative flexibility in machines and in the mechanicus, and early modern audiences expected mechanical performances to correspond to other means of conceiving both nature and society. In other words, mechanical performances became relevant to early modern audiences when they corresponded with, for example, organic conceptions of society as a body, patriarchal metaphors of the state as a house or a family, or theocratic means of understanding the state through analogy to the order of God's creation or the church as the body of Christ.

Mechanical practitioners and philosophers presented audiences with interpretations of nature as made up of infinitely small moving and interacting cogs and springs, not unlike the clocks and machines constructed by a mechanicus. This mechanical world was presented as in balance, in ways that corresponded to classical views of nature. In mechanical physico-theology, God was interpreted as the divine artificer and overseer, an ingenious being whose mechanical creations were infinitely more complex than those of any human mechanicus, and who guaranteed the running of his contraption.[40] Also the small world of the human body could be interpreted in mechanical terms. Iatromechanics made the body intelligible as a mechanical assemblage of related structures. Because the body was given mechanical properties, society – traditionally seen as an organic "body politic" – in turn could be interpreted as a machine–body consisting of subjects who, like cogs, worked together for a coherent whole.[41]

---

[39] Otto Mayr: *Authority, liberty & automatic machinery in early modern Europe* (Baltimore, 1986), compare 122–36 and 139–47.

[40] On mechanical metaphors and clock metaphors in the so-called scientific revolution, see Carolyn Merchant: *The death of nature. Women, ecology, and the scientific revolution* (San Francisco, 1980), 99–126; Mayr: *Authority, liberty & automatic machinery in early modern Europe*, 54–101; Shapin: *The scientific revolution*, 30–46; Dear: *The intelligibility of nature*, 15–38. For a discussion of mechanical metaphors in a Swedish context, see Dunér: *Världsmaskinen*; Dunér: *The natural philosophy of Emanuel Swedenborg*; See also Dunér: *Tankemaskinen*.

[41] On iatromechanics in the writings of William Harvey and René Descartes, see Thomas Fuchs: *The mechanization of the heart. Harvey and Descartes* (Rochester, N.Y, 2001), 115–96; See also Thomas Wright: *William Harvey. A life in circulation* (Oxford, 2012), 207. For a discussion of the relationship between natural and scientific order in seventeenth-century Europe, see Shapin and Schaffer: *Leviathan and the air-pump*; Peter Dear: "Mysteries of state, mysteries of nature. Authority, knowledge and expertise in the 17th Century", in Sheila Jasanoff (ed): *States of knowledge. The co-production of science and social order* (London, 2004), 206–24. On mechanics as a means of making society intelligible, see Mayr: *Authority, liberty & automatic machinery in early modern Europe*, 102–21; Simon Schaffer: "Enlightened Automata", in William Clark, Jan Golinski, and Simon Schaffer (eds): *The sciences in enlightened Europe* (Chicago, 1999), 126–65. For the use of mechanical metaphors in the Swedish political community, see Pasi Ihalainen: "New visions for the future. Bodi-

We are thus presented with a number of powerful mixed metaphors: of a social machine–body, of God as a creator of the perfect artifice of nature, and of machines as bodies and social orders crafted in matter. Put differently: mechanics and machinery mediated between understandings of change and order in both nature and society. By describing human bodies, societies and nature as products of mechanical work, these phenomena could all be understood through each other, because they were all objects of human or divine agency.[42] By means of these mixed metaphors, mechanical frameworks of understanding became integrated into other ways of interpreting the world (be they organic, patriarchal or theocratic). In early modern Sweden, these layers of interpretations formed a complex stack that could be used to legitimise a hierarchical society.[43] The mechanicus was a man who could convincingly present these mechanical visions of bringing order to the world. Whether he built machines, presented plans for educating young men in their image, or dreamt up ways to perfect the hierarchies of the early modern state, he was relevant to his contemporaries in the ways he could relate his mechanics to audiences' expectations.

Making trusted and virtuous men

Early moderns trusted the dreams of mechanici because these dreams were embedded in other contemporary networks of meaning. When mechanical practitioners performed in ways that were relevant to their time, they attained resources and authority to carry out their visions. Consequently, issues of trust and authority are at the heart of this thesis. Over the past decades, historians of science have taken an interest in early modern practices of asserting authority to define what was true, good and useful. These studies, which have approached knowledge making as a part of various early modern European cultures and states, provide a number of differing frameworks of trust and authority of relevance to this study of the mechanicus of the early modern Swedish state.

---

ly and mechanical conceptions of the political community in eighteenth-century Sweden", in Petri Karonen (ed): *Hopes and fears for the future in early modern Sweden, 1500–1800* (Helsinki, 2009), 315–40; Dunér: *Tankemaskinen*, 274.

[42] The concept of mediating machines were introduced by Norton Wise in his "Mediating machines", *Science in context* 2:1 (1988), 77–113. Though this term, Wise highlights both conceptual mediation, between "a network of concepts from political economy and a similar network in natural philosophy" and *methodological* mediation, where machines mediate between "the truth of theoretical knowledge in its utility and the utility of practical knowledge in its truth."; ibid., 77. I use this term in both these senses. Early modern machines were both a means of mediating between a social and natural order, as well as a means of mediating between seventeenth- and eighteenth epistemological concepts.

[43] Merchant has made a similar point, arguing that "The organic and mechanical philosophies of nature cannot, therefore, be viewed as strict dichotomies", Merchant: *The death of nature*, 103.

In *A social history of truth* (1994), Shapin examines "the great civility", an elite culture in seventeenth-century England in which credibility was identified in the social status of individual actors. According to Shapin, "the moral economy of premodern society located truth within the practical performances of everyday social order." By contrast, "the modern condition permits no such effective resort to familiarity and personal virtue in deciding upon the truth or falsity of knowledge-claims."[44] The gentleman was the quintessential truth-teller of early modern England and actors who were trusted invoked a notion of gentility. Because he was seen as free by virtue of his economic independence, the gentleman was identified as a truth-teller in contrast to those who "labored under constraint."[45] The recognition that the establishing of a specific form of knowledge also required the cultivation of a certain kind of knower and his findings, forms an important basis for this thesis. However, when studying the mechanicus of the early modern Swedish state, it is important to be sensitive to the specific cultural connotations given to mechanical work in this part of Northern Europe. In particular, it is necessary to recognise that few subjects of the Swedish realm had the financial resources of an English gentleman. In contrast to England, where trust was accredited to free-acting gentlemen, one finds in Sweden a system of authority, of a poor absolutist state, where submission and hard work were crucial for identification as a trustworthy man.

When discussing what was true and useful, one should thus be wary of providing too broad explanations of how actors established authority. Practices of trust and authority were part of specific cultures that varied over both time and place.[46] Several studies have shown alternative social frameworks of trust contemporary to "the great civility" of seventeenth-century England. For example, Jay Smith has pointed out how by accepting "Habermas's dichotomy between the state and the public sphere of private persons," many studies of the French *ancien régime* ignore how this absolute monarchy was "responsible for creating and gradually transforming an abstract public space."[47] Whereas Shapin focuses on how English gentlemen

---

[44] Steven Shapin: *A social history of truth. Civility and science in seventeenth-century England* (Chicago, 1994), quotes from pages 36 and 410–11; see also Steven Shapin: "The invisible technician", *American scientist* 77:6 (1989), 562; Jan Golinski: "The care of the self and the masculine birth of science", *History of science* 40 (2002), 125–45.

[45] Shapin: *A social history of truth*, 43.

[46] Peter Dear has made a similar call for historical specificity. He argues that the relationship between knowledge makers and the state is not a given. Such a relationship depends on which form of knowledge we study, and which state. Dear proposes that early modern authoritative knowledge could recast "the product of scientific activity as another kind of *order*, not social but natural"; Dear: "Mysteries of state, mysteries of nature", 206.

[47] Jay M. Smith: *The culture of merit. Nobility, royal service, and the making of absolute monarchy in France, 1600–1789* (Ann Arbor, 1996), 267. Compare to Shapin's account of the English Royal Society in *A social history of truth*, 120. Though a *royal* society obviously was related to the English state, it was not an integrated part of the state administration. In comparison, the organisation of the French *l'Académie de science*, for example, was closer linked to the government than its English counter-

gained trust by acting in a society of equals, Smith shows a French culture in which trust was given to actors working within the state apparatus.[48] The role of the state in early modern Sweden bears more similarities to France than to Britain. Because of their relative poverty and the general scarcity of resources, Swedish mechanical practitioners could not attain independence through private fortunes.[49] However, one should not equate the early modern Swedish state with the French *ancien régime*. The war-ridden early eighteenth-century Swedish absolutism had but a fraction of the economic resources of the French state, and, if we wish to understand the role of mechanical practitioners in early modern Sweden, it is important to see their relational performances, enacted in the state, as more than means to an end, or ways to gain material benefits.

Such relationships to powerful actors in the state were central to early modern men of science, as shown in Mario Biagioli's analysis of the court of the Medici. Compared with Shapin, who describes the Royal Society as a collective structure of relative peers, Biagioli describes the self-fashioning of Galileo in a court structured as a pyramid of patrons and clients.[50] By fashioning himself in relation to the anticipations of patrons, Galileo Galilei could act within this structure of patronage in order to further his own career, increase his status and legitimise his knowledge claims.[51] These asymmetric relationships carry many similarities to the ones that I am studying. However, whereas Biagioli primarily studies the strategic nature of these relationships, as a means of making knowledge in an early modern court culture, I am interested in how these relationships formed the mechanical practitioners themselves.

A number of recent studies have shown how early modern natural philosophy and mathematics were not only perceived as means of making

---

part. See for example: Daniel Roche: *La France des Lumières* (Paris, 1993), 461; see also Charles Coulston Gillispie: *Science and polity in France at the end of the old regime* (Princeton NJ, 1980), 79–81.

[48] Of course, the systems of trust in England and France should not be seen as complete opposites. Being seen as affiliated with the state and public interest was pivotal to the entrepreneurs and experimental philosophers of the English constitutional monarchy, e.g. in Stewart: *Rise of public science*, 16.

[49] For a discussion of early modern economy and trust in Sweden, see Klas Nyberg: "'Jag existerar endast genom att äga kredit'. Tillit, kreditvärdighet och finansiella nätverk i 1700-talets och det tidiga 1800-talets Stockholm", in Mats Berglund (ed): *Sakta vi gå genom stan. Stadshistoriska studier tillägnade Lars Nilsson den 31/5 2005* (Stockholm, 2005), 184–211.

[50] Mario Biagioli: *Galileo, courtier. The practice of science in the culture of absolutism* (Chicago, 1993), see esp. 4–5, 25–30. See also Mario Biagioli: "The social status of Italian mathematicians, 1450–1600", *History of Science* 27:1 (1989), 41–95. In his use of the concept of "self-fashioning", Biagioli follows studies of early modern history and literature. See for example Stephen Greenblatt: *Renaissance self-fashioning. From More to Shakespeare* (Chicago, 2005 [1980]). But while Greenblath focuses on the relationship between the expressions of cultural codes in literary works, Biagioli sees self-fashioning as an act in relation to a social structure of clients and patrons. For a similar study of structures of patrons and clients in a Swedish context, see Hjalmar Fors: *Mutual favours. The social and scientific practice of eighteenth-century Swedish chemistry* (Uppsala, 2003).

[51] Biagioli: *Galileo, courtier*, 99.

knowledge, but also as an exercise that transforms the knowing subject. Sorana Corneanu has shown how the works of Francis Bacon, Robert Boyle and John Locke related to a longer early modern tradition of *cultura* and *medicina animi*, to which the philosophical study of nature became a regimen of the mind. She argues that what would later be known as "objectivity", in the seventeenth century, would be "understood as virtuous dispositions acquired by disciplines meant to transform the 'temper' of the philosophers' minds." As such, the goal of natural inquiry was not solely to make knowledge, but also to exercise the mind of the inquirer; beyond offering a route to understanding nature, natural philosophy was further a means to craft oneself into a virtuous subject. Thus, Corneanu shows that the possibilities that early modern actors saw in experimental philosophy were not necessarily the same as those that modern historians ascribe to these activities.[52] Similarly, Matthew L. Jones has shown how René Descartes, Blaise Pascal and Gottfried Wilhelm Leibniz viewed mathematics, and especially geometry, as a spiritual exercise that could "help one live the good life."[53] Like these authors, I want to understand early modern mechanics and mathematics as self-forming exercises. However, my main focus is not how famous philosophers or mathematicians discussed mechanics, or other forms of mathematics, as spiritual exercises. Instead, I am interested in how men shaped themselves into useful and virtuous subjects by performing early modern mechanics in relation to each other.

To seventeenth- and eighteenth-century audiences, the terms "usefulness" and "virtue" carried different connotations from the ones they do today. In early modern economic discourse, the concepts of virtue and usefulness were intertwined, so that the one could not be without the other. A useful man acted in the interest of the *publicum* (i.e., of the state or the sovereign) as a contributing yet subordinated member of the state. Such a useful man was also a virtuous man, defined by a set of virtues such as diligence and concord. Virtue and usefulness were not only states of mind but also part of actors' performances.[54] As such, they should both be seen as relational concepts. Audiences identified virtue and usefulness in an actor's performance, and actors aligned their performances with what they anticipated the audience's expectations of a virtuous and useful man to be. No man was born virtuous: the ability to act virtuously and usefully needed to

---

[52] Corneanu: *Regimens of the mind*, 10–11. Stephen Gaukrouger has made similar arguments about Bacon's philosophy in *Francis Bacon and the transformation of early-modern philosophy* (Cambridge, 2001); Gaukroger: "The persona of the natural philosopher".

[53] Matthew L. Jones: *The good life in the scientific revolution. Descartes, Pascal, Leibniz, and the cultivation of virtue* (Chicago, 2006), 2.

[54] Leif Runefelt: *Hushållningens dygder. Affektlära, hushållningslära och ekonomiskt tänkande under svensk stormaktstid* (Stockholm, 2001), 88–9; Leif Runefelt: *Dygden som välståndets grund. Dygd, nytta och egennytta i frihetstidens ekonomiska tänkande* (Stockholm, 2005), 9.

be cultivated.[55] A virtuous and useful man was a man who inhabited a specific position in relation to his contemporaries, and the process of claiming this position can be seen as a form of enculturation.[56] As seen, for example, in the quote from Mört at the beginning of this chapter, many early modern pedagogical works presented mathematics as an exercise though which such a virtuous and useful subject could be made.[57] By studying the mechanicus, it is possible to understand this educational process, involving the performances of mechanical techniques, which shaped men according to the formal and informal norms of the early modern Swedish state.

The cameral mechanicus of the Swedish state

This thesis consists of four chapters, each of which presents a new context for understanding the mechanicus and mechanics as self-forming exercises. What holds these cases together is the way they all examine how mechanics was a means of becoming a subjugated member of the political order in early modern Sweden. So, what were these frameworks of trust and authority, of which the mechanicus was a part?

In the Swedish empire of the seventeenth and during the two first decades of the eighteenth centuries, the absolute monarch – who answered only to God – guaranteed the operation of the political order. During this period, Sweden was a major regional power in the Baltic Sea. After the end of the Great Northern War in 1721, Sweden had lost most of its eastern possessions, except for Finland. In the aftermath of the war, Sweden underwent drastic governmental changes, and in the 1720s it became a constitutional monarchy with a strong parliament. However, there was much continuity between absolutist and constitutional government: the estates of the parliament can be said to have replaced the former absolutist monarch, and to have established a "sovereignty of the Estates."[58] Also, the state administra-

---

[55] Leif Runefelt has pointed out how to early modern authors, virtue was supposed to be internalised through continuous exercise; Runefelt: *Hushållningens dygder*, 88.

[56] I approach education not as separated from the making of knowledge, but as a necessary and integrated part of it. See Harry M. Collins: "Learning through enculturation", in Angus Gellatly, Don Rogers, and John A. Sloboda (eds): *Cognition and social worlds* (Oxford, 1989), 205–15; Harry M. Collins: *Changing order. Replication and induction in scientific practice* (Chicago, 1992), 159–160.

[57] Similar views of mathematics, as a virtuous exercise, can be seen in other parts of Europe. For example, as shown by Kelly J. Whitmer, in the orphanage in Halle, mathematical sciences were considered "technologies of discernment" and mathematical exercises would teach children to reconcile opposing positions; Kelly J. Whitmer: "Eclecticism and the technologies of discernment in pietist pedagogy", *Journal of the history of ideas* 70:4 (2009), 545–67; Kelly J. Whitmer: *The Halle Orphanage as scientific community. Observation, eclecticism and pietism in the early enlightenment* (Chicago, 2015). Whitmer has kindly provided me with parts of the manuscript of this recent publication, which has been helpful in understanding the role of Halle eclecticism for the Swedish mechanicus.

[58] Michael Roberts: *The Age of Liberty. Sweden 1719–1772* (Cambridge, 1986), 62.

tion, or the state apparatus, that had developed during the Swedish imperial experience, remained and grew during the 1700s.

The core provinces of early modern Sweden lay along the shores of Lake Mälaren, by the West coast of the Baltic Sea. Stockholm, the capital of the realm, contained The Royal Palace as well as the state administration. The administration, carrying out the will of the sovereign, whether it was a monarch or the estates, was formed around what historical actors recognised as the norms and practices of "cameralism". These cameral norms and practices had a long history, ranging back to 1495 when the Holy Roman Emperor Maximilian I (1459–1519) had gathered his main administrative servants under a *Kammerkollegium*. The term derives from the Latin *camera*, denoting an institution of the court subjugated directly to the sovereign. By the seventeenth century, most German states and the states in Scandinavia had a fiscal chamber [*Kammer*], which managed the affairs of the sovereign.[59] Soon the state chamber of the Swedish empire developed into a number of specialised administrative bureaus [*kollegier*], focused on providing resources for the realm through, for example, mining, commerce and war.[60]

The concept of *œconomy* was central to early modern cameralism. It was a framework of economics – shaped by a mechanistic view of the world as well as the Greek concept of the household – according to which art, the study of nature and the financial matters of the state were intertwined in

---

[59] Wakefield: *The disordered police state*, 16–17. Pascale Laborier: "Les sciences camérales, prolégomènes à toute bureaucratie future ou parades pour gibiers de potence?", in Pascale Laborier et al. (eds): *Les sciences camérales. Activités pratiques et histoire des dispositifs publics* (Paris, 2011), 13. On the relation between the Swedish state administration and the sovereign, see Bo Lindberg: *Den antika skevheten. Politiska ord och begrepp i det tidig-moderna Sverige* (Stockholm, 2006), 12.
Following Andre Wakefield, in the ensuing text I use the adjective "German" to denote states that are part of the Holy German Empire. As Wakefield points out, the sentence "the states of the Holy Roman Empire" is a mouthful, and while these lands should not be identified as identical to present day Germany, historical actors of these states often identified themselves as "German". This was also usually how migrants from these states in the Swedish Empire were identified, as well as translations of books in the German language. Compare to Wakefield: *The disordered police state*, 20–1.

[60] David Gaunt: *Utbildning till statens tjänst. En kollektivbiografi av stormaktstidens hovrättsauskultanter* (Uppsala, 1975), 31–71; Maria Cavallin: *I kungens och folkets tjänst. Synen på den svenske ämbetsmannen 1750–1780* (Gothenburg, 2003), 83–93; Fors: *The limits of matter*, 46–7. Compare to the broader European picture in Laborier: "Les sciences camérales, prolégomènes à toute bureaucratie future ou parades pour gibiers de potence?", 13.
Following Hjalmar Fors, I translate *kollegium* into "bureau"; Fors: *The limits of matter*, 14–15. Some works on the Swedish administration of the seventeenth and eighteenth centuries have discussed these administrative bodies as "boards", e.g. Lindqvist: *Technology on trial*, 95. To talk of these bodies as boards highlights "the organisational collegial structure at the top level", where all matters were discussed and decided by a board. Others have made a more literal translation, and translated *kollegium* into "college", e.g. Fabian Persson: *Servants of fortune. The Swedish court between 1598 and 1721* (Lund, 1999), iii–v. As pointed out by Fors, this was also a common translation in the seventeenth and eighteenth centuries. By discussing these parts of the state administration as bureaus, it is possible to understand them as larger administrative bodies and highlight the integrated nature of their administrative, knowledge making and educational efforts; Fors: *The limits of matter*, 14–15.

ways that differed from the more narrowly defined economics of today.[61] Twentieth-century historians generally integrated œconomy into histories of the rise of the modern state, or declared it the ancestor of twentieth-century disciplines such as economy, sociology, political science or public administration.[62] As in the case of mechanics, in order to establish such relationships, the past has been interpreted through modern economic concepts and controversies. A number of more recent studies have instead shown how cameralism and œconomy were parts of historical political orders. The cameralists of the German and Scandinavian states worked in this intersection between knowledge of nature and political economy. The hands and eyes of the monarch, they oversaw the production and trade of the realm. As part of this endeavour, they both aimed to make the state intelligible to the sovereign, and to legitimise the sovereign's authority in the eyes of the inhabitants of the realm.[63] Andre Wakefield has discussed these cameralists

---

[61] Marc Raeff: "The well-ordered police state and the development of modernity in seventeenth- and eighteenth-century Europe. An attempt at a comparative approach", *The American historical review* 80:5 (1975), 1229. On German cameralism, see ibid.; Wakefield: *The disordered police state*, esp. 1–25. On French political economy, see Laborier: "Les sciences camérales, prolégomènes à toute bureaucratie future ou parades pour gibiers de potence?". On cameralism in Sweden, see Stellan Dahlgren: "Karl XI:s envälde - kameralistisk absolutism?", *Makt & vardag* (Stockholm, 1993), 115–32; Mårten Snickare: *Enväldets riter. Kungliga fester och ceremonier i gestaltning av Nicodemus Tessin den yngre* (Stockholm, 1999), 24. In England, a public science partly separated from state matters developed from the late seventeenth century. However, also there practical mechanics was intimately linked to state affairs as well as to providence. See Stewart: *Rise of public science*. For a general overview of cameralism and mercantilism in seventeenth- and eighteenth-century Europe, see Lars Magnusson: *Mercantilism. The shaping of an economic language* (London, 1994), 174–203. The interconnections of œconomy and natural philosophy in early modern Europe are explored in the collection by Margaret Schabas and Neil De Marchi (eds): *Oeconomies in the age of Newton* (Durham NC, 2004). For an overview, see Margaret Schabas and Neil De Marchi: "Introduction", in Schabas and De Marchi (eds): *Oeconomies in the age of Newton*, 1–13. For a discussion of the relationship between œconomy and natural history in Sweden, and especially in the works of Linnaeus, see Lisbet Rausing: "Underwriting the oeconomy. Linnaeus on nature and mind", in Schabas and De Marchi (eds): *Oeconomies in the age of Newton*, 173–203. The most in-depth history of the connections between natural philosophy, œconomy and mechanics in eighteenth-century Scandinavia can be found in Dan Christensen's study of technology and culture in Denmark–Norway 1750–1850, *Det moderne projekt. Teknik & kultur i Danmark-Norge 1750–(1814)–1850* (Copenhagen, 1996).

[62] Wakefield: *The disordered police state*, 5. The role of order and reform in Northern-European economic discourse, and the relationship between œconomy as a field of knowledge and administrative practice has been discussed by a number of scholars on German Cameralism. See for example Keith Tribe: "Cameralism and the science of government", *The journal of modern history* 56:2 (1984), 263–84; Keith Tribe: *Governing economy. The reformation of German economic discourse, 1750–1840* (Cambridge, 1988); Wakefield: *The disordered police state*.

[63] Alix Cooper: "'The possibilities of the land'. The inventory of 'natural riches' in the early modern German territories", in Schabas and De Marchi (eds): *Oeconomies in the age of Newton*, 131. Wakefield has pointed out a prevalent but unhappy habit in the historiography of German cameralism of separating "cameralists of the book" from "cameralists of the bureau"; Andre Wakefield: "Books, bureaus, and the historiography of cameralism", *European journal of law and economics* 19:3 (2005), 311. Hjalmar Fors has shown the importance of cameralism for understanding the epistemological position of mining officials in the early modern Swedish state; see Fors: *The limits of matter*, 101–3.

as "the alchemists of Enlightenment." They promised to find new sources of affluence in everyday places: "in local fields, forests, mines and manufactories."[64] Inquiries into art and nature were a standard ingredient in their recipe for domestic wealth. In the cameralists' line of work, matters of nature and œconomy were separated by "very permeable boundaries".[65] Consequently, the bureaus housed men skilled not only in œconomy, but also in, for example, chemistry and natural history. Also, and central to this thesis, many of these cameralists took an interest in mathematics.

The ways in which knowledge of nature and state utility intersected in early modern Sweden has been examined by a number of historians, most recently by Fors.[66] However, already nineteenth- and twentieth-century scholars have pointed out how, in the eighteenth century, natural history, chemistry and mathematics (including mechanics) were seen as means, first of supporting the military endeavours of the Swedish empire and then of restoring the dwindling empire to its former power.[67] The ideology behind this mixed natural inquiry and economy has often been discussed as "utilism", a term used in Swedish eighteenth-century historiography that lacks a good English counterpart. Generally, in Swedish historical research, this term has come to designate cameral ideologies similar to those found in contemporary German states.[68] While nineteenth- and twentieth-century historians primarily focused on the role of useful knowledge in the universities and the Royal Swedish Academy of Sciences [*Kungliga vetenskapsakademien*], more recently scholars have focused on the role of the state bureaus and the military as sites of knowledge production.[69] These scholars have also

---

[64] Wakefield: "Books, bureaus, and the historiography of cameralism", 319.

[65] Citation from Schabas and De Marchi: "Introduction", 9. On cameralism and natural history as a means of finding wealth in the quotidian and local in eighteenth-century Germany and Scandinavia, see Cooper: "The possibilities of the land", 134; Alix Cooper: *Inventing the indigenous. Local knowledge and natural history in early modern Europe* (Cambridge, 2007), 22, 99. On Cameralism and natural philosophy in the Baltic realms, see Lisbet Rausing: "Daedalus Hyperboreus. Baltic natural history and mineralogy in the Enlightenment", in Clark, Golinski, and Schaffer (eds): *The sciences in enlightened Europe*, 397–8. See also Lisbet Rausing: *Linnaeus. Nature and nation* (Cambridge MA, 1999), 3–6.

[66] Fors: *The limits of matter.*

[67] Claes Annerstedt: *Upsala universitets historia* 3 vols. (Uppsala, 1908); Bengt Hildebrand: *Kungl. Svenska Vetenskapsakademien. Förhistoria, grundläggning och första organisation* vol. 1 (Stockholm, 1939); Sten Lindroth: *Kungl. Svenska vetenskapsakademiens historia 1739–1818. 1:1, Tiden intill Wargentins död (1783)* (Stockholm, 1967); Sten Lindroth: *Svensk lärdomshistoria. Frihetstiden* (Stockholm, 1997 [1978]).

[68] For a discussion of the translation of this term, see Fors: *Mutual favours*, 22. As pointed out by Runefelt, "utilism" is a concept invented by modern historians who have applied it to the eighteenth-century as part of a long narrative of the de-moralisation of the economy. See Runefelt: *Dygden som välståndets grund*, 14. In this thesis, this term only appears in relation to the work of previous historians.

[69] For a general picture of the relationship between natural knowledge and the state bureaucracy, see Sven Widmalm: "Instituting science in Sweden", in Roy Porter and Mikuláš Teich (eds): *The scientific revolution in national context* (Cambridge, 1992), 240–62; Arne Hessenbruch: "The spread of precision measurement in Scandinavia 1660–1800", in Kostas Gavroglu (ed): *The sciences in the*

recognised that the Baltic region should be studied as a unit, and have related this national historiography to the role of useful knowledge in the bureaus of German states.[70]

Together, all these studies have shown how natural philosophy and mathematics were integrated parts of the early modern Swedish state, and that actors who carried out these activities were generally civil servants. However, these studies have not examined the expectations that contemporary audiences had of these civil servants, and how they in turn made themselves as they made knowledge. By approaching mechanics as performances by which men made themselves into subjects of the state, I want to understand mathematics and mechanics not only as tools of the state but also as exercises by which civil servants were made.

The cameralists and mechanical practitioners should thus not be approached as two separate groups: a large part of the œconomical writers in the Swedish eighteenth-century state were also mechanical projectors and vice versa. All the actors of the first half of the 1700s that I study in this thesis inhabited this position: they presented cameral treatises in which mechanics was the vehicle for imagining a future orderly society *and* they designed machines motivated by œconomical arguments. If cameralists were the alchemists of enlightenment, it seems as if mechanics and mathematics were key ingredients in their brew. For these mechanist–cameralist authors, a mechanistically conceived state became an argument for the usefulness of machines, and the power of machines became the foundation of a mechanistic vision of social order. Their recipe was hardly a new one: the promises

European periphery during the Enlightenment (Dordrecht, 1999), 179–224. On the roles of mathematical sciences and astronomy in early modern Swedish state, especially in regards to mapmaking, see Sven Widmalm: *Mellan kartan och verkligheten. Geodesi och kartläggning, 1695–1860* (Uppsala, 1990). On mechanics in the Bureau of Mines [*Bergskollegium*], see Lindqvist: *Technology on trial*. On mechanics in the seventeenth-century state, see Dahl: *Svensk ingenjörskonst under stormaktstiden*. Natural history and œconomy has been discussed by Rausing: *Linnaeus*; Hanna Hodacs and Kenneth Nyberg: *Naturalhistoria på resande fot. Om att forska, undervisa och göra karriär i 1700-talets Sverige* (Stockholm, 2007). The role of chemistry in the Swedish Bureau of Mines is discussed in Fors: *Mutual favours*; Hjalmar Fors: "Kemi, paracelsism och mekanisk filosofi. Bergskollegium och Uppsala cirka 1680–1770", *Lychnos* (2007), 165–98; Anders Lundgren: "Gruvor och kemi under 1700-talet i Sverige. Nytta och vetenskap", *Lychnos* (2008), 13. Thomas Kaiserfeld has used the case of saltpetre boiling as a case for analysing the role of scientific knowledge in the Swedish military state of the 1700s; Thomas Kaiserfeld: *Krigets salt. Salpetersjudning som politik och vetenskap i den svenska skattemilitära staten under frihetstid och gustaviansk tid* (Lund, 2009). On the rise and decline of useful natural knowledge in eighteenth-century Swedish state, see Karin Johannisson: "Naturvetenskap på reträtt. En diskussion om naturvetenskapens status under svenskt 1700-tal", *Lychnos* (1979), 107–54.
[70] Lisbeth Rausing has argued that "Baltic Enlightened science" should be seen as a "Germano-Scandinavian one", Rausing: "Daedalus Hyperboreus", 389; Similarly, Hjalmar Fors has discussed the role of German immigrants in Swedish mining, and the German-Swedish area as a unit in Fors: *The limits of matter*, 103–9, 153. For the view of Sweden as a country of learning among eighteenth-century German scholars, see Mathias Persson: *Det nära främmande. Svensk lärdom och politik i en tysk tidning 1753–1792* (Uppsala, 2009), 76–146.

made by mechanici were "related to a much older belief in the fabricator of machines as a magician or sorcerer of some kind, capable of creating mechanical life."[71] However, although this was an old mix, it was still related to vivid dreams of the future. Through this powerful tincture – of intertwined œconomy, mechanics and virtue ethics – the mechanicus promised order and prosperity to a state obsessed with its own stability.

Few historical studies have focused on why eighteenth-century European societies, in which the dreams of mechanical projectors were so seldom realised, repeatedly trusted in these projectors' promises and rewarded their performance financially and symbolically.[72] Instead, histories of eighteenth-century mechanics have generally followed long narratives and been less interested in historical inconsistencies than in explaining the rise of certain modern phenomena. Whereas the successful good mechanicus in these studies becomes an agent of change, pivotal in introducing modern modes of mechanical production, the failed bad mechanicus at most becomes a futile dreamer and an irrelevant visionary. In the coming chapters, I integrate these two characters by focusing on early modern mechanics as a process of self-formation. It is the relational performances that made and unmade these men, and the promise that they carried to audiences in the early modern Swedish state, which form the heart of this thesis.

Sources and translations

In the following chapters, I approach the mechanicus using different categories of sources and by complementing the methodology presented here with additional theoretical perspectives. Therefore, each chapter begins with a more detailed discussion of the sources it is based on, as well as its theoretical approach. The overwhelming majority of the sources are written in Swedish (some are also written in Latin, German or French) and they are found in publications and archive documents, written by actors with diverse relationships to the early modern Swedish state. Because published English translations only exist for a few of these source documents, the translations used in this book are generally my own; when this is not the case, I indicate it in a footnote. Accompanying my own translations, I provide the original Swedish quotations in the footnotes.[73] In the few cases when translations have already been published, I do not provide the original citations.

---

[71] Sawday: *Engines of the imagination*, 191. See also J. Peter Zetterberg: "The mistaking of 'the mathematicks' for magic in Tudor and Stuart England", *The sixteenth century journal* 11:1 (1980), 83–97.

[72] Some recent studies have started to take these questions seriously. See for example Sawday: *Engines of the imagination*; Wakefield: "Leibniz and the wind machines".

[73] In the original quotes, I have aimed to preserve the original historical spelling, changing only archaic typography (by, for example, consistently changing dashes into commas). When quoting

Swedish has changed radically over the past centuries, compared with relatively more conservative languages such as English or French. To make eighteenth-century Swedish intelligible to modern native readers is thus an act of translation in itself. To translate eighteenth-century Swedish into modern English further complicates this process. By preserving the original quotes in my footnotes, I want to make this complex process of translation transparent. In some cases – when the original Swedish wording is of vital consequence for the interpretation, and for translations of central concepts – I put the original wording in brackets after the translation. Moreover, a glossary of frequently reoccurring terms can be found at the end of this book.

Some terms are more complicated to translate than others. The actors did not discuss mechanics in terms of technology, or in any equivalent terms, and thus it is relatively easy to abstain from using this term. However, the use of words translatable into "science" is a somewhat more complicated matter. Eighteenth-century Swedes regularly referred to mathematics and other fields of knowledge as *wettenskaper* or *wetenskaper*. These terms are roughly translatable to the Latin *scientia*, denoting both certain systematic knowledge and the fields of knowledge from which it derived. Depending on the context in which the term was used, I translate *wettenskap* as either "knowledge" or, when the term implies a field of systematic knowledge making, as "science". Thus, when I use the term "science", this should not be identified as equivalent to modern science, but instead as comparable to *scientia*.

Chapter outline

Mechanical practitioners can be found in numerous parts of the early modern Swedish state apparatus and the cases I present here are not, nor do I want them to be, a complete description of early modern mechanics. Instead, they are means of studying how mechanical practitioners made themselves into subjects of the political order of their time. They all ask questions of how mechanics was a self-forming exercise, and how it was a means of performing both submission and superiority in the civil service. Each chapter thus focuses on sources that illuminate these relationships. Rather than seeking a single institutional or geographical home for the mechanics, I delineate his location in an eighteenth-century culture of superiority and submission. It would be possible to study other aspects of the mechanicus – for example, how mechanics was performed when actors travelled between

---

works printed in German type, I represent sections printed in Antiqua for emphasis by using italics. I have generally been restrictive with adding italics for emphasis in my quotes. Hence, italics in quotes are from the original, unless stated otherwise.

states, or the expectations of, and demands on, mechanical performances in the military endeavours of the Swedish state. However, this is not my focus here, which is reflected in my choice of cases.

This thesis has a thematic structure and is divided into four chapters, in addition to this introduction and the conclusion. Each of these four chapters studies the mechanicus through encounters between specific actors and audiences. The chapters gradually progress from the general to the specific. The chapter succeeding this introduction discusses mechanical coming-of-age narratives, as imagined by mature mechanical practitioners. It argues that mechanics was presented as a rite of passage – consisting of crafts and mathematics – through which a mechanically apt boy became a mechanicus with the virtues expected of a manly subject of a religious and political order. Then, in Chapter 3, I narrow the scope to a community of civil servants, in the Swedish Bureau of Mines [*Bergskollegium*], formed around mathematics and mechanics. Here I argue that in the Bureau, over the first half of the 1700s, mathematics developed from being the concern of a small group of mining mechanics to being a technique by which young men could become part of a community that saw the world in geometric and mechanical terms. In this chapter, I thus show how mathematical and mechanical exercise carried a symbolic role in a specific community of civil servants. In Chapter 4, I study the role of mechanics, mathematics and œconomy in the correspondence of the mechanicus Christopher Polhammar and the absolute king Karl XII as well as of men in their vicinity. Through this correspondence, I discuss how two early modern personas – the monarch and the mechanicus – were performed in relation to each other. Finally, I present a case study of a single actor – the today virtually unknown mechanical practitioner Anders Gabriel Duhre – who first succeeded and then failed to present himself as a virtuous mechanicus to the parliament of the Swedish constitutional monarchy of the 1720s and 1730s. This case ties together many of the themes of the previous chapters, and analyses how a mechanical practitioner could succeed, and fail, in performing as a virtuous mechanicus in relation to various audiences in the state.

2. Mechanics – a passage into manhood

In this chapter, imagination takes centre stage. Specifically, it focuses on how mechanics was imagined as a rite of passage. In prefaces, speeches and texts on education one finds reoccurring coming-of-age narratives, which revolved around boys' performances of mechanics in relation to older experienced men. Such texts imagined the life of a mechanicus to be a *via media* of a useful man, a middle ground in between a number of other personas: of the man of Latin, the pure mathematician and of the craftsman, to name but a few. These narratives are the objects of study of this chapter, and my aim is to understand how they were both expressions and constitutive of the mechanicus. Or in other words: how these narratives followed and formulated expectations on how to enter a form of useful manhood by performing mechanical techniques such as mathematics and crafts.

In 1716 the mechanical practitioner, Christopher Polhem (known as Polhammar until his ennoblement in December the same year), published a short treatise for the education of boys and young men in mathematics with the somewhat impervious title of *The second foundation of wisdom. For the glory of youth, the usefulness of manhood and the pleasure of old age.* In a short introduction, the author justified the book's title: he explained that manhood was about maturity, and mathematics and mechanics were means of becoming mature. Because the young by nature tended primarily to reflect on things that gave them "glory and honour, instead of long-term usefulness", it would not hurt to "spice up the latter with the former, in the book itself as well as its name". In this way, boys would be guided into "the usefulness of manhood", and would learn things that could earn them their daily bread. Polhem also wished that his book would give pleasure to the old who took delight in "teaching and educating their child in everything related to their lawful work and well-being". For him, mathematics was one of two pillars of a virtuous and useful life, the other being language.[1]

Thirty-two years later, Lars Laurel – a philosopher and librarian at the University of Lund – commemorated the late mechanicus Mårten Triewald in the newly established Royal Swedish Academy of Sciences. In his speech,

---

[1] "beröm och heder, än nytta i lengden"; "så wel i sielfwa wercket som nambnet bekryddar thet samma til hog af thet ena"; "Mandoms nytta"; "lära och underwisa sin barn i alt hwad som til theras lofliga näring och welferd lender"; Christopher Polhem: *Wishetens andra grundwahl til ungdoms prydnad mandoms nytto och ålderdoms nöje; lempadt för ungdomen efter theras tiltagande åhr, uti dagliga lexor fördelt* (Uppsala, 1716), "En liten Förklaring öfwer Titlen", unpaginated.

Laurel went off at a short historical tangent about how mechanics had recently become a manly activity: "When our great king Karl XII lived, who never rested from battles or military accomplishments [*kämpedater*]," he argued, "we saw how the able young were mostly inclined to only weapons and armour." Everything else, "though not less useful," was considered "not to be manly practices [*karle-förer*]". Instead, according to Laurel, such techniques were considered the "crafts of weaklings and women." [2] But, Laurel contrasted:

> Under the government of our current beloved king, the conditions are again reversed. Alongside weapons, scholarly arts and handicrafts have come in fashion, so that there is almost no one who does not engage with them in some way. Thus, one may say that these peacetime deeds have never carried the honour and glory among us that they now claim. [3]

While Polhem and Laurel both discussed mathematics and mechanics in relation to manhood and manliness, they did so from two different perspectives. Whereas Polhem treated manhood as the part of life between youth and old age, and mechanics as a means of maturity, Laurel argued that mechanics had become a way to express a set of attributes that distinguished manly subjects from women, children and effeminate men. In eighteenth-century Sweden, the term "manhood" carried both these connotations; manhood was an age in which esteemed manly qualities could be identified, as well as the qualities that identified someone as having reached this age. The two earlier examples can thus be seen as two variations on a common narrative, in which mechanical and mathematical education was related to the coming of age of a useful man.

In this chapter, I study how mechanical practitioners imagined the coming of age of such a useful man through mechanical exercise using two complementary sources. The first category – where we can place Polhem's preface – consists of fourteen prefaces, and introductions, of textbooks on mathematics and mechanics, written in Swedish and published roughly between 1700 and 1750. Borrowing from Gerard Genette such texts, together with book titles, can be seen as the "peritext" of a book (as in "periphery"), which frames its content. As such, introductions and prefaces are generally thresholds of interpretation. Thus, coming-of-age narratives of mechanics found in prefaces are especially interesting for understanding the meanings ascribed to a mechanicus, because they have been put in a textual space that

---

[2] "När vår uti strid och Kämpedater sig aldrig hvilande *store Konung Kung* CARL den XII lefde"; "såge vi den raske ungdomens hog mäst stå til vapen och brynja allena"; "ehuru icke mindre nödigt"; "ej vara Karle-förer"; "veklingars och qvinne-slögder."; Lars Laurel: *Åminnelse-tal öfver capitaine mechanichus vid fortificationen … Mårten Trievald, hållet på store riddar-hus salen, den 23 decemb. 1747* (Stockholm, 1748), 5.

[3] "Under vår nu hulde Konungs regering är det åter så omvändt, at jämte vapen boklige konster och handslögder vunnet det tycke, at ingen nästan är, som icke på något sätt sig dermed inblandar, så at man väl kann säga, at desse fredsbragder aldrig varet hos os i den heder och ära, som de nu kommet til."; ibid.

aims to establish a shared understanding between author and reader.[4] I supplement these prefaces with fourteen other, more general educational, texts that also discuss mathematics. The second category consists of fourteen commemorative speeches from the Swedish Academy of Sciences – such as the one by Laurel – that contain narratives of coming of age in which mechanics and mathematics feature. These texts are from the mid 1700s, and they look back at and reminisce about boyhoods in the first half of the century.

The two categories provide complementary narratives of mechanics and maturity: one looking forward, expressing an mature man's gaze on boys and their path in life, and one looking backwards, in which men turn their eyes back onto themselves in trying to understand the product of the boys they once had been. These narratives of boys, childhood and manhood – prevalent in eighteenth-century texts on mechanics and mathematics – have attracted little attention from historians. Instead, as discussed in the previous chapter, such sources have been used to understand early modern technology. Likewise, autobiographies and commemorative speeches of mechanical practitioners have been read as descriptive sources for biographies of exceptional individuals.[5] Thus, by approaching these texts as narratives of manhood and ageing, I wish to highlight a completely different aspect of mechanics in early eighteenth-century Sweden – that is, mechanics as an exercise imagined to shape boys into a certain form of man, defined in relation to a political order.

In the following section, after a brief discussion of the meanings ascribed to manhood and the ways in which it structured the early modern Swedish state, I turn to the sources where mechanical practitioners discussed the signs of a mechanically apt boy and where they imagined ideal relationships between apt boys and experienced men such as themselves. Then, in the next two sections, I analyse the role of mathematics and crafts in these relationships and in the imagined coming of age of a mechanicus – in other words, how a range of authors imagined that performance of these two sets of techniques fostered a boy into a specific form of manhood.

---

[4] Gerard Genette: *Paratexts. Thresholds of interpretation* (Cambridge, 1997), 4–5.
[5] See e.g. Samuel Bring, who described Christopher Polhem as a "pioneer [*banbrytare*]" who aimed to "create a Swedish industry" (in original: "skapa en svensk industri"); Bring: "Bidrag till Christopher Polhems lefnadsteckning", 107. This narrative is still prevalent in popular discourse on eighteenth-century mechanical practitioners. For example, Michael Lindgren has described how "Polhem in many respects was a pioneer" (in original: "Christopher Polhem var i många avseenden en föregångare"); Lindgren: *Christopher Polhems testamente*, 281.

Even more striking, a recent exhibition about Polhem at The National Museum of Science and Technology in Stockholm carried the illustrative title "Polhem – Back to the Future"; National Museum of Science and Technology, Stockholm: "Christopher Polhem. Tillbaka till framtiden" <http://www.tekniskamuseet.se/1/2604.html> [accessed 27 March 2015].

Age, manhood and the pious subject

What meanings were ascribed to a boy's passage into manhood in the eighteenth century? I approach manhood as a distinction of gender, and I see gender as both relational and organisational – in other words, as something that inhabits social structures, relationships and the imagination.[6] A number of scholars of the past decades have shown how male gender identity was (and is) performed in relation to women as well as to other men.[7] In eighteenth-century Europe, performances of manhood were bounded by a variety of intersecting distinctions: not only of gender but also of social status and of age. As pointed out by Anthony Fletcher: "if we are to see something of what being a man meant in past societies, we need to loosen the hold of modern concepts of gender", while we remain sensitive to the fact that "gender was often cross-cut with other differentials of status."[8] In the eighteenth century, manhood was a mandate of authority in relation to women, children and men of lower social status, as well as a constant and stable age in-between disorderly youth and degenerative old age.[9] For a boy to enter manhood, he needed to be raised and educated according to his social status. While mothers were in charge of the education of younger children, fathers and tutors were supposed to chaperone older boys into manhood. These older men were to guide and exercise fatherly authority over the younger boy; by performing a variety of manly activities together, the older man should encourage esteemed attributes in his son or student.[10]

Manhood as pious submission

In a patriarchy such as early modern Sweden, this fatherly role had implications that extended beyond the relationship between man and boy. As argued by Michael Roper and John Tosh, the study of patriarchy "provides a way of integrating the individual and structural dimensions of male dominance." Following Max Weber, they see patriarchy as various systems of organisational domination structured around the concept of "father-rule", instead of as a general structure of power.[11] In this Weberian sense, early

---

[6] For a similar methodological approach, see Michael Roper and John Tosh (eds): *Manful assertions. Masculinities in Britain since 1800* (London, 1991), 11.

[7] See for example Raewyn Connell: *Masculinities* (Cambridge, 2005), 37; Philip Carter: *Men and the emergence of polite society. Britain 1660–1800* (Harlow, 2001), 5. See also Michael Roper and John Tosh: "Introduction", in Roper and Tosh (eds): *Manful assertions*, 4.

[8] Anthony Fletcher: *Gender, sex and subordination in England 1500–1800* (New Haven, 1995), 89.

[9] On manhood and age in the eighteenth-century, see Alexandra Shepard: *Meanings of manhood in early modern England* (Oxford, 2003), 21.

[10] Anthony Fletcher: *Growing up in England. The experience of childhood, 1600–1914* (New Haven, 2008), 129, 144. See also John Tosh: "What should historians do with masculinity? Reflections on nineteenth-century Britain", *History workshop* 38 (1994), 181.

[11] Roper and Tosh (eds): "Introduction", 9–10. For a discussion of Weber on patriarchy as the organisational domination of father-rule, see Malcolm Waters: "Patriarchy and viriarchy. An

modern Sweden was profoundly patriarchal. The absolute monarchy of late seventeenth- and early eighteenth-century Sweden was legitimated by Lutheran theocratic ideology that merged the power of God, of the king and of fathers.[12] God, who had created the world, was its lawful sovereign and, as such, he had delegated his power according to an order where fathers ruled over the women, servants and children of their house [*hushåll*]. As mankind multiplied, the divine Lord had aggregated some of this fatherly power in the monarch, who would guarantee an orderly society under his rule.[13] In this theocratic order, loyal subjects were identified through a hierarchy of virtues. In spite of being strongly associated with the male norm, the same hierarchy governed the expected performance of all subjects, woman or man. Still, the relative importance of the virtues, and *how* these virtues were supposed to be performed, differed according to social distinctions such as gender, social status and age.[14]

At the top of this hierarchy of virtues was *pietas*, or piety. The modern concept of piety, with its mainly religious connotations, does not fully capture the meaning of *pietas* in seventeenth- and early eighteenth-century Europe. At this time, piety denoted the expected performance of a loyal subject in a state where religion and politics were intertwined. Besides

---

exploration and reconstruction of concepts of masculine domination", *Sociology* 23:2 (1989), 195–6; William Clark: *Academic charisma and the origins of the research university* (Chicago, 2006), 9.

[12] For a discussion of the similar French absolutism, see Gauvin: *Habits of knowledge*, 168–9. Gauvin points out how "Religious morality and complete obedience to a father figure – family patriarch, king, God – became effective means of enforcing a measure of civil order. Mind and body were no longer private property of beings in ancien régime France. Individuals became social bodies, inseperable from the royal and religious authorities of the kingdom."; ibid. The same holds true for Sweden during its absolutist regime, Lindberg: *Den antika skevheten*, 98. For a general discussion of gender as "a primary way of signifying relationships of power"; see Joan W. Scott: "Gender. A useful category of historical analysis", *The American historical review* 91:5 (1986), 1069.

[13] Nils Ekedahl: *Det svenska Israel. Myt och retorik i Haquin Spegels predikokonst* (Uppsala, 1999), 84–5; Snickare: *Enväldets riter*, 23–5; Kekke Stadin: *Stånd och genus i stormaktstidens Sverige* (Lund, 2004), 16–21. Compare to the account of manhood and the patriarch in the Holy Roman Empire of the same time in Lyndal Roper: *Oedipus and the Devil. Witchcraft, sexuality and religion in early modern Europe* (London, 1994), 46. The constitutional monarchy post 1720 did not make any significant break with this Lutheran ideology. (On the patriarchal order of Swedish Lutheran orthodoxy post-1720, see Carola Nordbäck: *Samvetets röst. Om mötet mellan luthersk ortodoxi och konservativ pietism i 1720-talets Sverige* (Umeå, 2004), 399. While the monarch was stripped of his direct relationship with the Lord, the emerging concept of citizenship was still ordered around fatherly authority ordained by God, which was distributed through society along intersecting lines of social distinction. See Jonas Nordin: *Frihetstidens monarki. Konungamakt och offentlighet i 1700-talets Sverige* (Stockholm, 2009), 83–7; Karin Sennefelt: *Politikens hjärta. Medborgarskap, manlighet och plats i frihetstidens Stockholm* (Stockholm, 2011), 278.

[14] Kekke Stadin: "Att vara god eller att göra sin plikt? Dygd och genus i 1600-talets Sverige", in Janne Backlund (ed): *Historiska etyder* (Uppsala, 1997), 228–30; Åsa Karlsson: "En man i statens tjänst. Den politiska elitens manlighetsideal under det karolinska enväldet 1680–1718", in Anne Marie Berggren (ed): *Manligt och omanligt i ett historiskt perspektiv* (Uppsala, 1999), 116; Stefan Rimm: *Vältalighet och mannafostran. Retorikutbildningen i svenska skolor och gymnasier 1724–1807* (Uppsala, 2011), 42–3.

encompassing submission to God, the performance of piety involved the fulfilment of duties to one's parents and the fatherland: it was the merging of religiosity with fidelity to the state and to one's superiors. These two parts of a pious performance were prerequisites for each other: a good Christian was an obedient subject and an obedient subject was also submissive to God and to the church. As pointed out by Carola Nordbäck, at this time Lutheranism and Swedish patriotism were a unit. A trusted patriot was a pious man, who diligently worked for the public good and who realised that this involved maintaining the hegemonic position that the Lutheran church held in the realm. Men who failed to perform in the interest of the public were "*monstra moralia*" (i.e., "immoral freaks").[15] Piety thus summarised the expectations of how a submissive subject was supposed to act according to their position in the theocratic order.[16] In this order, an act of pious submission gave authority and credibility to both the submissive and superior part of a relationship; hence, authority was not a zero-sum game.[17] As shown by Kristiina Savin, in a world governed by the will of God there existed a correspondence between the universe and human action. Success could be explained by the fact that the divine Lord loved virtue.[18] In such a world, the proposition that pious submission enabled a man to act was not only a social, but also an ontological fact. These conflicting demands of authority and submission meant that discussions and performances of pious manhood were anything but homogeneous. Instead, the claims of mandate inherent in manhood – that distinguished useful men from *monstra moralia* – were anxiety-ridden performances. These performances placed a subject in a patriarchal hierarchy of submission and superiority, a place that constantly needed to be checked, maintained and reconsolidated.[19]

The useful knowledge of a theocratic state

The patriarchal order also permeated early modern knowledge production. During the Swedish absolute monarchy of the seventeenth and early eighteenth centuries, scholars were subject to a religious as well as a political orthodoxy according to which the autocracy, as well as the Lutheran teach-

---

[15] Nordbäck: *Samvetets röst*, 134–6, quotes from page 136.

[16] Kurt Johannesson: "Om furstars och aristokraters dygder. Reflexioner kring Johannes Schefferus Memorabilia", in Sten Åke Nilsson and Margareta Ramsay (eds): *1600-talets ansikte* (Nyhamnsläge, 1997), 316; Snickare: *Enväldets riter*, 24; Janne Lindqvist: *Dygdens förvandlingar. Begreppet dygd i tillfällestryck till handelsmän före 1770* (Uppsala, 2002), 59–60.

[17] Karin Sennefelt has made a similar point, that, in mid-eighteenth-century Sweden, submission hardly eliminated the ability to act politically; Sennefelt: *Politikens hjärta*, 280–1.

[18] Kristiina Savin: *Fortunas klädnader. Lycka, olycka och risk i det tidigmoderna Sverige* (Lund, 2011), 50–1.

[19] Compare to the performance of English gentlemanliness, discussed in Michèle Cohen: "'Manners' make the man. Politeness, chivalry, and the construction of masculinity, 1750–1830", *Journal of British studies* 44:2 (2005), 315.

ings, were above criticism by decree. This order continued into the constitutional monarchy of the eighteenth century, when an academic dissertation was not allowed to contradict Lutheran orthodoxy, good morals or the form of the government.[20] In the early eighteenth century, Lutheran orthodoxy defined itself in relation to radical forms of pietism, which came to Sweden in the forms of heretics and heretic works primarily originating in the German states. But the relationships between pietism and Lutheran orthodoxy was not straightforward. Whereas the church officially saw radical pietism as a threat, the educational and utilitarian aspects of pietism were seen favourably by many civil servants and clergymen, especially such who had studied and travelled abroad.[21] In the Swedish universities of the first decades of the eighteenth century, the eclecticism of pietist scholars from Halle – such as Johann Franz Buddeaus (1667–1729) and Christian Thomasius (1655–1728) – became an important tool for clergymen and university professors who wished to reconcile conservative pietism with Lutheran orthodoxy.

The clergyman and philosophy professor of Lund University, Andreas Rydelius (1671–1738), was central to the process of reconciliation.[22] Philosophically, Rydelius was strongly shaped by Cartesianism, and he was also a prominent figure in the Swedish Lutheran church (in 1734 he became Bishop of Lund). Still, in his work *Necessary exercises of reason* (1718–20), he displayed a tolerant position towards conservative pietism and his work also showed an affiliation with the Halle eclectics.[23] Rydelius' book was encouraged by the current king Karl XII, and can be seen as an authoritative text on the social epistemology of the theocracy. As such, it shows how powerful audiences in the Swedish state imagined knowledge to weave together the minds and bodies of knowers in the theocratic order. Also, it was used and referred to as an authoritative text up until the mid 1700s.[24] Rydelius

---

[20] Claes Annerstedt: *Upsala universitets historia. D. 2, 1655–1718, 2, Universitetets organisation och verksamhet* (Uppsala, 1909), 131.

[21] Nordbäck: *Samvetets röst*, 403.

[22] Tore Frängsmyr: *Wolffianismens genombrott i Uppsala. Frihetstida universitetsfilosofi till 1700-talets mitt* (Uppsala, 1972), 42–3.

[23] Jonathan Israel has presented Rydelius as "an admirerer of Buddeus as well as Descartes and Andala," who waged war on "Wolffians, Lockeans, and Spinozisits alike until his death in 1738"; Jonathan I. Israel: *Radical enlightenment. Philosophy and the making of modernity 1650–1750* (Oxford, 2001), 560, 561. Israel also presents Rydelius as a persiting defender the "Cartesian ascendency in Sweden-Finland" of the early eighteenth century; ibid., 561. Still, in his *Necessary exercises* Rydelius comes forth as much less of a dogmatic, especially when presenting mathematical and physical literature. The case for Rydelius as an ecclectic is also made by Sven Wermlund: *Sensus internus och sensus intimus. Studier i Andreas Rydelius filosofi* (Stockholm, 1944), 279–84; Karl Gustaf Almquist: *Andreas Rydelius' etiska åskådning* (Lund, 1955), 187. Similarly, Teddy Brunius has described Rydelius as a mediator between Decartianism and Lockean empiricism, Teddy Brunius: *Andreas Rydelius och hans filosofi* (Lund, 1958), 12.

[24] References to Rydelius can for example be found in Weidler and Mört (tran.): *En klar och tydelig genstig*, preface; Gustav Ruder: *Anledning til snille-walet, eller Ungdomens snille-gåfwors och naturliga böjelsers bepröfwande, wal och anförande, til the wetenskaper, konster, ämbeten och syslor, med hwilka hwar och en kan wäl och beqwämligen tjena Gudi och fädernes landet* (Stockholm, 1737), 114–23; Eric Eklund:

defined useful knowledge as having three results: it improved the will of the knower, it led to worldly improvement and it extended the glory of God. Of the three, he considered the glory of God to be the ultimate end. However, the most useful knowledge was that which led to "the improvement of the mind of its possessor", because such knowledge was what eventually would honour God the most.[25] In other words, useful knowledge was what made the knower a pious part of a patriarchal order. Useful experience and skill were not external to the man who possessed them. Instead, useful knowledge reflected back upon and shaped the knowing subject into a man who worked for the glory of God and his fatherland; a useful man was thus a pious man.[26]

In his study of the coronation ceremony of Queen Ulrika Eleanora in 1680, Mårten Snickare analyses a staging of the early modern theocracy in the form of a *mons pietatis* – a mountain of piety. In this monument, work and useful knowledge intersected. The theme of the monument was the process of going from childhood into adult life and becoming a pious subject. The mountain of piety was shaped like a pyramid. It alluded to the architecture of antiquity, while it, at the same time, was a geometric allegory of a stable society.[27] At its right-hand side sat the boys of the orphanage, to the left sat the girls, and at the top rested a woman, personifying piety. Under the boys stood four girls, personifying expected virtues of a manly subject: *theoria* (carrying a folded paper and a pair of compasses pointed upwards), *praxis* (with a plummet and a pair of compasses pointed downwards), diligence (holding a clock) and concord (carrying three hearts bound together).[28] Interestingly, on the mountain of piety, all male virtues except concord were symbolised by what at the time would have been considered mathematical instruments: plummets, compasses and clocks. Also, the epistemic categories of *theoria* and *praxis* were placed together with more expected virtues of an early modern subject. While *theoria* and *praxis* might not commonly have been presented as expected virtues of a subject, their place on the mountain of piety again shows how early modern knowledge was inherently moral and that education was useful in relation to the virtuous

---

Upfostrings-läran, som wisar sätt och medel til ungdomens rätta skiötsel och underwisning (Stockholm, 1746), 251. See also Frängsmyr: *Wolffianismens genombrott i Uppsala*, 40–2.

[25] "ägarens förbättring til sitt Sinne"; Andreas Rydelius: *Nödige förnufftz öfningar för all slags studerande ungdom, som wil hafwa sunda tankar, och fälla ett billigt omdöme om de högste och wichtigste ting i wärlden, hwilket i naturlig måtta bör wara deras yppersta ändamål, ehwad för stånds wahl de hälst willja giöra* vol. 1 (Linköping, 1718), unpaginated, § 5.

[26] Rydelius view of the practical and useful could thus also be related to the medicina-cultura animi, as discussed in Corneanu: *Regimens of the mind*, 58.

[27] On the Tessin's use of antique, and especially Roman, architectural ideals, see Snickare: *Enväldets riter*, 183–4.

[28] Ibid., 51–4; Mårten Snickare: "Shaping the ritual space. Nicodemus Tessin the younger and Swedish royal ceremonies", in Allan Ellenius (ed): *Baroque dreams. Art and vision in Sweden in the era of greatness* (Uppsala, 2003), 134–5.

subject it fostered.[29] But then what kind of virtues were they? At a time when natural and moral concerns were separated by very permeable boundaries, *theoria* and *praxis* can be said to have been both *moral* and *epistemic* virtues. That is, they were both codes of a pious subject, and of performing as a trustworthy maker of knowledge about art and nature.[30] In the following, I will argue that in early eighteenth-century Sweden, a boy's passage into manhood involved the acquisition of a specific set of virtues, especially concord, diligence, *theoria*, and *praxis*. A manly subject was to exhibit both authority and pious submission, and he should be both productive and orderly. The coming-of-age narratives of mechanics, discussed in the following section, imagined one way into such pious manhood. At the heart of these narratives were the patriarchal relationships of a theocratic state, in the form of intergenerational relationships formed through mutual mechanical interests.

Imagined intergenerational relationships

It seems as if mechanical practitioners were obsessed with the issue of how best to craft boys into their own manly image. They repeatedly wrote texts in which they imagined how their guidance would lead young boys into useful manhood. In speeches, they verbalised their expectations of younger generations, discussed how best to observe mechanical potential in a child and contemplated how to lead a boy into a mechanical way of life. In an address to the Royal Swedish Academy of Sciences in 1750, the instrument maker Daniel Ekström argued that when mechanically apt boys were "nursed from the age of ten or fifteen, they gain far greater skill, than those who start at a later age". Still, it was difficult to identify aptitude in boys that young. The best bet, according to Ekström, was to search for boys who themselves, through "their own natural impulse", exhibited the possession of "some reflection, enterprise, diligence and precision in small crafts". According to Ekström, such boys were not interested in socialising with boys of their own age. Instead, they wished "to be alone or in the company of older people." For him, mechanical aptitude was expressed through a bond that reached across generations. As Ekström saw it, the starting point of a

---

[29] For a discussion of the most common virtues discussed in early modern Swedish sermons, see Stadin: "Att vara god eller att göra sin plikt?", 229. The relationship between virtue and knowledge-making in early modern Europe has been discussed by a number of historians of science. See for example Shapin: *A social history of truth*, 68, 410–11; Jones: *The good life in the scientific revolution*, 8–9; Corneanu: *Regimens of the mind*, 10–11, 34–7. For more longe durée studies of virtue in natural inquiery, see Lorraine Daston and Peter Gailson: *Objectivity* (New York, 2007).

[30] The term epistemic virtues has been proposed by Lorraine Daston and Peter Galison to designate "historically specific ways of investigating and picturing nature"; Daston and Gailson: *Objectivity*, 28. On the relation between moral and natural concerns in the seventeenth and eighteenth centuries, see Schabas and De Marchi: "Introduction", 9.

mechanical way of life was a reciprocal act of identification: the young boy would recognise the mature man from whom he could learn the mechanical techniques to which he was inclined, and the man would recognise the child's potential. Ekström imagined that such a boy, undisturbed by the distractions offered by children his own age, would respect and learn from a skilful older man, and that the two would interact using the mechanical techniques that they both loved.[31]

The identification of aptitude

Boys' practice of mathematics and crafts – the two major sets of techniques that constituted performance of mechanics – were common topics in commemorative speeches of the Royal Swedish Academy of Sciences of the mid-eighteenth century. The Academy's speeches often ascribed the prominence of former members to the fact that their aptitude had been identified early on by their elders. Anders Johan von Höpken described how for the young Anders Celsius mathematics had been a child's game. "Already in childhood, he amused himself with *geometric figures*," a game that "at his age meant much and early revealed his inclination and aptitude."[32] Similarly, Olof von Dalin recounted how as a boy Pehr Elvius (the younger) was "slow in learning Greek, because neither reason nor imagination found pleasure in a mere matter of memory." Instead, he had expressed an interest in drawing. After having drawn the royal castle of Drottningholm "better than one could have expected from his age," his way of life was set out for him: "everyone pointed him to *mathematics*".[33] It was the duty of older men to identify and lead mechanically inclined boys into manhood. Commemo-

---

[31] "upammade ifrån 10 à 15 års åldren, vinna de långt större färdighet, än de, som börja vid senare ålder"; "egen naturs drift"; "någon eftertanka, tiltagsne, flitige och granlaga uti små slögder"; "deras hog och böjelse"; "äro häldre ensamne eller uti de äldres sälskap."; Daniel Ekström: *Tal, om järn-förädlingens nytta och vårdande* (Stockholm, 1750), 24.

[32] "Redan i barndomen roade hans sig med *Geometriske figurer*,"; "för hans år mycket betydde och förrådde tidigt des hug och snille."; Anders Johan von Höpken: *Åminnelse-tal öfver astronomiæ professoren Anders Celsius efter kongl. vetenskaps academiens befalning hållit i stora riddarhus-salen d. 27 novemb. 1745.* (Stockholm, 1745), 7.

[33] "trög, at lära Græska [sic]; ty hvarken omdöme eller inbillnings-gåfva funno i en blott minnessak sit fulla nöje."; "fullkomligare, än man väntat af hans år,"; "dömdes han af alla til *Mathematiquen*"; Olof von Dalin: *Åminnelse-tal öfver kongl. vetenskaps academiens medlem och secreterare, herr Pehr Elvius* (Stockholm, 1750), 4. Similar narratives, of how aptitude had been identified already in childhood, can be seen in a number of commemorative speeches. See e.g., Herman Diedrich Spöring: *Johannis Moræi … äre-minne, i auditorio illustri på riddarehuset efter kongl. wetenskaps academiens befallning uprättadt* (Stockholm, 1743), 13–14; Samuel Klingenstierna: *Åminnelse-tal öfver kongl. vetensk. academiens framledne ledamot, commerce-rådet Christopher Polhem, på kongl. vetenskaps academiens vägnar hållit i stora riddarhus-salen* (Stockholm, 1753), 3–4; Pehr Wilhelm Wargentin: *Åminnelse-tal öfver kongl. observatoren och kongl. vetenskaps academiens ledamot, herr Olof Hiorter, efter kongl. vetenskaps academiens befallning, hållit i stora riddarhus-salen d. 18 april, 1751* (Stockholm, 1751), 6–9; Jacob Henrik Mörk: *Åminnelse-tal öfver slotts-byggmästaren och kongl. vetenskaps academiens ledamot, herr Claes Eliander, hållit den 27 februarii, 1756* (Stockholm, 1756), 7, 9–10.

rating Daniel Ekström, Pehr Wargentin pointed out how apt boys were often born who "easily learn, what others must acquire through much education and effort," but that "such noble seeds seldom become ripe fruit: however fertile, they nonetheless always require some early tending, or at least guidance and opportunity to develop their gifts."[34]

In another commemorative speech, given by Jacob Hendrik Mörk on the clergyman Nils Brelin, the central theme was this act of identifying a boy's aptitude. Mörk recommended each and every person to examine "what he by nature is meant as well as called for." Nature was an "artful mother, who makes it so that among many thousands of faces, one is seldom similar to the other in all aspects". In the same way, "during the creation of such a distinctive body", she "instilled quite different dispositions." She gave them "distinctive bodily benefits, and also diverse shares of the powers of the mind." Mörk warned that when we "choose the wrong path in life, the public will miss much benefit". Brelin had been such a man who had trod the wrong path. He had possessed "living imagination, a fortunate ability of invention, a lucid reflection and a profound reason," and Mörk pointed out that by using these powers of mind Brelin had "gone far in the clerical estate, so that he has been honoured with the doctor's hat." Still, Mörk dared to propose that Brelin would have been a better mathematician than clergyman. Brelin should have been given "a sufficient mathematical foundation" at an early age and an "insight into the mechanical laws of motion". Also, when young he should have "exercised the most prominent craft techniques [handlag]". Had Brelin received such an education, his inventions could have competed with "the quickest minds of Europe". Then, Brelin would have made notable contributions to "our private as well as public economy".[35]

---

[34] "lätt fatta, hvad andre genom mycken undervisning och möda måste inhämta,"; "sådana ädla frön komma sällan til mogen frukt: ehuru bördiga, fordra de likväl altid någon ans i början, eller åtminstone anledning och tilfälle at utvikla sina gåfvor."; Pehr Wilhelm Wargentin: *Åminnelse-tal öfver kongl. vetensk. academiens framledne ledamot, directeuren och mathematiska instrument-makaren, herr Daniel Ekström, hållit för kongl. vetenskaps academien, den 14 junii år 1758* (Stockholm, 1758), 4–5.

[35] "hvartil han af Naturen är ämnad och likasom kallad."; "konstrika Modren, som gör, at ibland så många tusende ansigten sällan et är det andra til alla delar likt"; "under en så åtskillig kropps skapnad nedlagt rätt så olika lynnen."; "åtskilligt kroppens förmoner, och äfven så olika låtter af sinnes gåfvor."; Jacob Henrik Mörk: *Åminnelse-tal öfver framledne theologie doctoren och kyrko-herden i Bolstad i Carlstads stift, herr Nils Brelin, på kongl. vetenskaps academiens vägnar, hållit uti stora riddare-hus salen* (Stockholm, 1754), 2. "misstage oss i valet af lefnads-vägar, saknar det allmänna mycket til sin förmon"; ibid., 3. "en lefvande inbillnings kraft, en lyckelig upfinnings-gåfva, en redig eftertanka och et djupsinnigt förstånd,"; "bragt sig i Kyrko-ståndet så högt, at han blifvit med Doctors-hatten hedrad."; "en tilräcklig Mathematisk underbyggnad"; "insigt i de Mechaniska rörelses lagar"; "inöfning i de förnämsta handa-lag"; "Europas qvickaste hufvuden"; "vår så allmänna som enskyldta hushållning"; ibid., 5.

Love of mechanics

Mechanical practitioners considered it up to them to guide apt boys on to the right path in life. Laurel presented the intergenerational relationships between experienced men and apt boys as filled with love, but this love could also be a problem if unchecked. Therefore, the self-love of boys needed to be contained. For the young, he warned, "it was difficult to know oneself". Gifted boys often believed that they "already possess [the knowledge], which they show good promise in," a premature opinion that they often received from careless tutors. According to Laurel, the act of identifying an apt young subject produced a form of "love" that permeated intergenerational relationships between boys and experienced men. By binding boys and men together, such love could make "old age insufficiently careful" and it could compel tutors "prematurely to give such beloved students too much". Thus, too warm feelings could lead to flattery and be detrimental to the boys' development: many apt boys had even "been ruined in this way."[36]

Laurel contrasted these misled boys to Triewald, the object of his speech, who as a child had displayed love for his elders and his fatherland as well as a deep knowledge of himself. In spite of his love for his father – presented as a *"Tubalcain* of our time" in reference to the biblical metalworker – Triewald had broken their natural patriarchal relationship by choosing not to take up his craft.[37] The young Triewald, Laurel suggested, had not left his father out of "a premature opinion of himself" or out of "contempt," but as an effect of the "natural gift and inclination, which the divine Lord has given him in order to be useful to his fatherland in an even more beneficial way."[38] For Laurel, intergenerational relationships between boys and men were built on a love that came from a mutual recognition. However, both parties needed to keep their emotions in check and to maintain the asymmetrical relationship that the difference in age prescribed. The young man's self-love should not mislead him to believe that he was his tutor's peer, and the tutor's love for the young man should not lead him to the same false conclusion. Instead, the boy should submit to experienced mature age, while it was the duty of the older and experienced man to identify those boys who had the aptitude and inclination necessary for them to

---

[36] "Det ät [sic] svårt att känna sig själv"; "redan besitja det, hvarom de gifva en god förhoppning," "kärlek"; "ej nog försigtig ålderdom"; "i förtid tilägnar sådane käre ämnen förmycket"; "blefvet så fördärfvade."; Laurel: *Åminnelse-tal öfver [...] Mårten Trievald*, 7. The commemorative speeches contain many concrete examples of intergenerational love and affection. See e.g. Spöring: *Johannis Moræi äre-minne*, 14; Wargentin: *Åminnelse-tal öfver [...] Olof Hiorter*, 12; Axel Fredrik Cronstedt: *Åminnelse-tal öfver framledne directeuren och kongl. vetensk. acad. ledamot Henric Theoph. Scheffer, på kongl. vetensk. acad. vägnar, hållit i stora riddarehus-salen, den 17. september 1760* (Stockholm, 1760), 9.
[37] "en vår tids *Tubalkain*"; Laurel: *Åminnelse-tal öfver [...] Mårten Trievald*, 3.
[38] "otidig tanka om sig sjelf"; "förakt,"; "den naturliga gåfva och drift, som den Högste honom gifvet til at på et ännu fördelaktigare vis vara Fäderneslandet nyttig."; ibid., 5.

prosper, and then to lead them into useful manhood for the benefit of their fatherland.

Such balanced love was the prerequisite of the mutual acts of identification on which these intergenerational relationships were imagined to rely. A mature mechanical practitioner, who wished to educate the young, should thus know the signs by which he could identify aptitude. Polhem recommended that children should be allowed to "run about in the house, and to potter about with all chores together with both parents and servants." Such bodily work would make the children "steady, diligent, quick and busy", but would also provide occasions for children's aptitude to be identified. Polhem also considered the performance of crafts a sign by which aptitude could be identified: "if then they seem to take pleasure in carving and crafting, that is a sign of a future *mechanicus*."[39] Likewise, according to Ekström, apt boys did not quickly tire of performing crafts. Their "inventiveness [*fintlighet*]" was revealed to the experienced observer through "the changes they wish to make [to the craft techniques]"; also, their "handiness [*händighet*]" and "imagination [*inbillnings-gåfva*]" were shown by "the likeness [of their results] to what they have decided to make." Besides crafts, Ekström imagined that mechanically inclined boys took an interest in mathematical arts. They liked "drawing, counting, geometry and to observe moving machines".[40]

It was not sufficient for a boy to show an interest in mathematics and crafts in order to be identified as a mechanicus: the passage from boy to mechanicus was also imagined to require a continuous exercise of these techniques. In their narratives, authors linked such exercise to the acquisition of virtues that defined a good manly subject: to diligence and concord as well as to the epistemic virtues of *theoria* and *praxis*. By analysing narratives where mathematics and crafts were part of the making of a useful man, it is possible to understand how early modern authors imagined these techniques as self-forming exercises carried out in intergenerational relationships between boys and men.

Mathematical concord

So far, I have presented the role of manhood and maturity in the formation of manly subjects of the early modern Swedish state. There, certain virtues –

---

[39] "löpa omkring i huset och stöka i alla syslor tillika med föräldrar och Tienstefolck."; "stadige, idoge, trefne och sysselsamme"; "Om då lusten skönjes hos dem, at tälja och slögda, så är det et tecken til en tilkommande *Mechanicus."*; Christopher Polhem: "Tankar om mekaniken", *Kungliga vetenskapsakademins handlingar* 1 (1740), 195.
[40] "de ändringar, de deruppå vilja hafva"; "yppar sig i likheten emot det de sjelfve sätta sig före."; "Rita, räkna, geometrie och at åskåda rörliga verk"; Ekström: *Tal, om järn-förädlingens nytta och vårdande*, 24.

such as diligence, concord, *theoria* and *praxis* – defined a manly subject, and mechanics and other practical mathematical arts were presented as parts of a process of maturity. In speeches and prefaces, one finds narratives of mechanics as a passage to manhood and of intergenerational relationships that formed both teachers and students. The authors of these texts imagined themselves pious men, who could hand down useful manhood to apt boys. In these coming-of-age narratives, the integrated exercise of mathematics and crafts was a central means of shaping a useful subject. Thus, the following sections analyse these self-forming exercises more closely. First, I examine mathematics as an exercise of fostering the reason and imagination required of a useful man.

When mature mathematicians discussed mathematics as a path to manhood, they often did so in relation to the classical curriculum of schools and academies. They presented mathematics as a means of education that was an alternative, or at least complementary, to Latin. Authors on mathematics warned that careless teachers in schools and academies – dominated by a classical curriculum – often failed to identify the potential of mechanically apt boys and consequently turned them away from their institutions. For example, Ekström argued that because these boys often lacked an "inclination for languages" they were frequently excluded from the schools and universities with a classical education taught in written and oral Latin.[41] At the same time, these authors clearly imagined mathematics and Latin to have similar roles in a boy's education. By understanding how mathematics was related to Latin, we can understand how, and why, mathematics was presented as an alternative or a replacement of a classical curriculum.

As pointed out by Walter Ong, because of its base in totally male communities, from medieval times onwards "Learned Latin" became a "sex-linked" language, "written and spoken only by males, learned outside the home in a tribal setting which was in effect a male puberty rite". Emerging from this rite of passage, which also involved studies in, for example, rhetoric and logic, was a scholarly man welcome in universities, in the correspondence of an international "republic of letters" and in the clerical estate.[42] Fletcher's observation that Latin in England 1600–1800 became "firmly installed as the male elite's secret language, [...] that could be displayed as a mark of learning, of superiority, of class and gender difference," also holds true for the Sweden of this time.[43] Bo Lindberg has argued that

---

[41] "lust för språk"; ibid.
[42] Walter J. Ong: *Orality and literacy. The technologizing of the word* (London, 1991), 113. On the role of Latin in the Swedish schools, see Wilhelm Sjöstrand: *Pedagogikens historia* vol. 3:1 (Lund, 1958), 136; Rimm: *Vältalighet och mannafostran*, 83.
[43] Fletcher: *Gender, sex and subordination in England 1500–1800*, 302. On Latin as an initiation into scholarly culture in eighteenth-century Sweden, see Rimm: *Vältalighet och mannafostran*, 244–6. For a discussion of classical education, and especially rhetoric, as power generating see Ann Öhrberg: *Samtalets retorik. Belevade kulturer och offentlig kommunikation i svenskt 1700-tal* (Höör, 2014), 30–6.

Latin and Swedish existed side by side in the ruling elite of eighteenth-century Sweden and that they can be seen as two parallel codes. One can roughly say that whereas Latin permeated the universities, the work of the civil service in Stockholm was mainly exercised in Swedish and its officials sought to cultivate the Swedish language, in order to put it on a par with Latin as well as other European languages.[44] Of course, the universities and the state were not separate from each other: most men of the state had been educated and formed in academia and as a consequence the bureaus in Stockholm had expectations of what such an education should entail. Like the cameralists of the German states, who strived to make universities – e.g., in Halle and Göttingen – into factories for educating civil servants, the cameralists in Stockholm imagined turning the universities of the realm into institutions that produced men for public positions – in the military or the civil administration – and who were well versed in juridical matters, rhetoric, œconomy, natural history and mathematics.[45]

Over the century, the curriculum of the schools was reformed slightly, to include a broadly defined natural history.[46] This constant attempt at reform was perhaps most clear in the policies in relation to the university in Uppsala, the largest university of the realm. In the seventeenth and eighteenth centuries, the bureaus in Stockholm pushed this traditionally clerically dominated institution in the direction of a practical œconomical academy.[47] At the same time, a number of mathematical works in Swedish appeared. At the end of the seventeenth century, most textbooks on mathematics were written in Latin. In the 1700s this situation slowly changed. The choice of language of these works was not arbitrary: there existed pedagogical motives for publishing in Swedish instead of Latin. In the preface to his Swedish translation of Euclid's *Elementa* (1744), Mårten Strömer argued that the young would not encounter useful knowledge "by the right time, as long as it is hidden from them in foreign languages".[48] Strömer was not

---

[44] Lindberg: *Den antika skevheten*, 26.

[45] On Swedish academia and state bureaucracy, see Gaunt: *Utbildning till statens tjänst*, esp. 31. The relationship between the bureaus and the universities was similar in many contemporary German states, see Clark: *Academic charisma and the origins of the research university*, 13.

[46] Sjöstrand: *Pedagogikens historia* vol. 3:1, 139–40, 147–50; Rimm: *Vältalighet och mannafostran*, 83–4.

[47] Wilhelm Sjöstrand: *Pedagogikens historia* vol. 2 (Lund, 1958), 232; Fors: *The limits of matter*, 84. For studies on knowledge making in the Swedish state administration, see the references in note 69 page 35. One expression of this conflict was the Cartesian controversies in Uppsala of the late seventeenth century, where Aristotelian and Cartesian theses were pitted against each other; Claes Annerstedt: *Upsala universitets historia. D. 2, 1655–1718, 1, Universitetets öden* (Uppsala, 1908), 91–101; Annerstedt: *Upsala universitets historia. D. 2, 1655–1718, 2, Universitetets organisation och verksamhet*, 310–13.

[48] "i rättan tid dertil, så länge den uti främmande språk för dem är undangiömd."; Euklides and Mårten Strömer (tran.): *Euclidis Elementa eller grundeliga inledning til geometrien, til riksens ungdoms tienst på svenska språket utgifven* (Uppsala, 1744), preface. Besides *Euclidis Elementa*, several textbooks discussed the importance of translating mathematical works into Swedish. See e.g., Anders Gabriel Duhre: *Första delen af en grundad geometria, bewijst uti de föreläsningar, som äro håldne på swänska*

alone in advocating early mathematical studies, nor was he the only one portraying Latin as an obstacle for teaching mathematics to boys. These arguments can be seen repeatedly in prefaces to mathematical textbooks.[49]

These prefaces did not present the relation between mathematics and Latin as one of simple opposites: instead, it was a relationship of difference and similarity. On the one hand, mathematics was presented as a competing, and superior, alternative to a classical education. In early modern Europe, mathematical studies were a manly rite of passage, and an exercise that fostered a gentle, honest and manageable subject.[50] Therefore, to translate mathematical works from Latin into Swedish was to facilitate the making of a mathematical man, serving the state. Like in the France of this time, in Sweden mechanics and mathematics were not only methods of knowledge making but also instruments of authority aimed at "fashioning a new ideal Man, one that could serve adequately both the state and *scientia*".[51] In this vein, the surveyor Eric Agner commenced his Swedish textbook *Arithmetica fractionum* (1710) with a poem from "the book to the reader" that presented mathematics as important "if you want to be a man". By the diligent study of fractions, Agner unassumingly poetised, "You become sensible, for others wise | and rightly reap glory as a prize."[52] If mathematically apt boys studied Latin and Greek before they approached mathematics, they would learn mathematics much later than if they could study it directly in their mother tongue. Consequently, their becoming a mature mathematical man of the state would require more time. Also, to untangle mathematics from Latin was a means of controlling boys' intergenerational relationships. If a mathematically apt boy studied Latin at a formative age, he risked being lured away from the mathematical life he was suitable for, to instead end up a scholar or a clergyman.

On the other hand, the texts presented mathematical exercises as in many ways similar to the language studies described by Ong: as a rite of passage that gave access to an all-male language. Mathematics, like Latin,

---

språket uppå kongl. fortifications contoiret i Stockholm (Stockholm, 1721), preface, unpaginated, last page; Fredrik Palmqvist: *Inledning til algebra. Första delen* (Stockholm, 1748), 6.

[49] In his translation of Johann Freidrich Weidler's discussed in the previous chapter, Johan Mört argued that it was important to introduce mathematics to boys early in life. Weidler and Mört (tran.): *En klar och tydelig genstig*, preface. Similarly, Anders Celsius proposed that mathematical studies should be commenced "already in the early years" (in original: "i de unga åren") in Anders Celsius: *Arithmetica eller Räkne-konst, grundeligen demonstrerad af Anders Celsius* (Uppsala, 1727), preface. A similar argument can also be found in Jacob Faggot's introduction to Pehr Elvius' translation of Clairaut's *Elements de géometrie*; Alexis Claude Clairaut and Pehr Elvius (tran.): *Inledning til geometrien. Af herr Clairaut* (Stockholm, 1744), "Läsare!", unpaginated.

[50] John Lewis Heilbron: *Geometry civilized. History, culture, and technique* (Oxford, 2000), 23–4.

[51] Gauvin: *Habits of knowledge*, 128. Compare to Pamela H. Smith's discussion of the identity of "the new philosopher" in Smith: *The body of the artisan*, 19–20.

[52] "Boken til Läsaren,"; "om tu wilt blij Man,"; "Så blir tu klook, för androm wijs / Och bär af snällom låf och prijs,"; Eric Agner: *Arithmetica fractionum. Thet är: räkne-konst vthi brutne-tahl* (Stockholm, 1710).

was a symbolic system separated from everyday life that carried connotations of exactitude and truth.[53] For authors of Swedish mathematical textbooks of the early eighteenth century, mathematics provided a new way of seeing. This view is similar to the one expressed by eclectic scholars in Halle. There, as shown by Kelly J. Whitmer, the visual aspects of mathematics made it considered a suitable exercise in learning to discern the world in a new way, and mathematics was considered a technique of discernment. Through mathematical exercises, students were believed to learn how to conciliate and judge between opposing philosophical positions.[54] Thus, although mathematical textbooks presented mathematics and Latin as alternative routes to manhood, these two rites of passages were not simple opposites in a struggle between modernity and antiquity. Instead, the classical education established a pattern, which the mathematical both adopted and criticised.

Like Latin, mathematics, and especially geometry, was interwoven with the political and religious order of early modern Europe. Because God was benevolent, he had provided means for the human soul to discern his providential design, and mathematics was one such means. For seventeenth-century Lutheran natural philosophers, a mechanically and geometrically conceived universe, with a passively contrived nature, became an argument for the radical sovereignty of God. According to Rydelius, mathematics gave opportunity for "deep and respectful thoughts about God and his hidden wisdom, which variably shows itself in the three principles of the great world: numbers, measurements and weights".[55] Through mathematics, and

---

[53] On Latin as a means of establishing objectivity, see Ong: *Orality and literacy. The technologizing of the word*, 113–14. Dunér has also argued that the use of mathematics also excluded unwanted participants from natural philosophical discussion, see Dunér: *Världsmaskinen*, 85.
[54] Whitmer: "Eclecticism and the technologies of discernment in pietist pedagogy"; 567; Whitmer: *The Halle Orphanage as Scientific Community*.
[55] "diupa och wördsamma tankar om GUD och hans fördolda wijshet, som föränderligen lyser uti de tre stora wärldens grundstycken, Tal, Mått och Wicht"; Rydelius: *Nödige förnufftz öfningar* vol. 1, § 12. Mört repeated Rydelius' discussion about God and mathematics verbatim in his preface to Weidler and Mört (tran.): *En klar och tydelig genstig*. For Swedish eighteenth-century mathematical accounts of God's creation, see for example Christopher Polhem: "[Att Gud alzmächtig ähr hela naturens uphof…]", in Axel Liljencrantz (ed): *Christopher Polhems efterlämnade skrifter* vol. 3 (Uppsala & Stockholm, 1952), 304–15; Anders Wahlström: *Ethica mathematica. Heller mathematisk sedelära* (Västerås, 1742). The relationship between early modern Lutheran theology and geometry has been analysed in Peter Barker and Bernard R. Goldstein: "Theological foundations of Kepler's astronomy", *Osiris* 16 (2001), 101–2; See also Richard S. Westfall: "The rise of science and the decline of orthodox Christianity. A study of Kepler, Descartes, and Newton", in David C. Lindberg and Ronald L. Numbers (eds): *God and nature. Historical essays on the encounter between Christianity and science* (Berkeley, 1986), 221–3. For a discussion of mechanism and Lutheranism, see Gary B. Deason: "Reformation theology and the mechanistic conception of nature", in Lindberg and Numbers (eds): *God and nature*, 170. Jones has discussed early modern geometry as a spiritual exercise, ultimately leading to a good life; see Jones: *The good life in the scientific revolution*, 15–53. For a discussion of this role of mathematics in a Swedish context, see Dunér: *Tankemaskinen*, 45–7. This argument, about mathematics as a religious exercise, of reason and consent,

again especially geometry, it was possible to perceive these hidden truths and relationships in nature. Thus, whereas Latin provided means of participating in theological discourse, mathematics was imagined as a way to interpret God's creation, or the book of nature. In the Swedish Lutheran theocracy – where the state was imagined to mirror the rational order of God's creation – authors could consequently plausibly argue that mathematics was a way to align oneself to a political order.[56] Mathematical studies would teach a boy to see nature and society as God and the sovereign intended them to be seen, and as such they were exercises that fostered a pious man. Mathematics was thus an exercise of pious discernment, by which a boy would learn to consent to his superiors and to align himself with the theocratic order.

An exercise of discernment

In the preface to his compilation of Anders Gabriel Duhre's mathematical lectures from 1718, Georg Brandt discussed mathematics as an exercise of reason using a metaphor of bodily practice. If carried out diligently, mathematical studies strengthened the reason and formed the mind of a boy:

> your reason is increasingly sharpened and improved by a diligent practice [of mathematics], so that you become much more skilled in work, which requires reason and thoughtfulness, than someone who has not [practised it]. This is not unlike the fact that he, who has trained his body though all sorts of *exercise* and practice, is much more agile and stronger than someone else who is equally gifted by nature but who has not improved this gift through good exercise.[57]

Celsius made a similar argument in the preface to his *Arithmetica* (1727) – a textbook that was widely used in Swedish schools and academies until the mid 1700s.[58] Because mathematics was an orderly form of knowledge, mathematical exercise would let boys "arrange their thoughts according to

---

is also made in Sverker Lundin: *Skolans matematik. En kritisk analys av den svenska skolmatematikens förhistoria, uppkomst och utveckling* (Uppsala, 2008), 139.

[56] Making this connection between mathematics and an authoritarian political order explicit, the English philosopher Thomas Hobbes' argued that the power of his Leviathan, the ideal absolute monarch, was analogous to the force of geometric inferences. See Shapin and Schaffer: *Leviathan and the air-pump*, 153.

[57] "ens förstånd genom flitig öfning der utinnan blifwer alt mer och mer skiärpat och förmerat, så at man til alla sådana syslor, som fordra förstånd och eftertanka, warder mycket skickeligare än en annan som eij sökt sitt förstånd med dylika eftertänksamma och wichtige saker at upbruka; äfwen som den hwilken har öfwat sin kropp uti allehanda *exercitier* och öfningar, är mycket wigare och starkare än en annan af lijka *naturens* förmåner som försummat dem igenom goda öfningar at föröka."; Georg Brandt: *En grundelig anledning til mathesin universalem och algebram, efter herr And. Gabr. Duhres håldne prælectioner sammanskrifven* (Stockholm, 1718), preface.

[58] Illustratively, in 1750 Fredric Palmqvist discussed Celsius' Arithmetics as a "masterwork" (in original:"mästarstycke"), and as "rather useful for those who wish to study mathematics" (in original: in original: "ganska tjenlig för dem, som antingen tänka at öfwa sig i *mathematiken*"); Fredrik Palmqvist: *Underwisning i räkne-konsten* (Stockholm, 1750), preface.

an order of reason". By mathematical exercise, "*students* would be trained from childhood to extend their powers of reasoning". Such boys would grow up "not to accept anything without proper foundation and *cause*, and to acquire a taste and to find pleasure in *demonstrations*." As an effect, students of mathematics would mature into men who could serve their fatherland by acquiring "a skilled reason" by which they could make solid and indisputable proofs. Through mathematical studies, boys could enter a realm of absolute certainty, where they could make truths that both they and others were obliged to accept.[59]

But how was a boy to enter this realm? Celsius pointed out that:

> a correct *logica*, or theory of reason, might provide rules and *maxims*; but even if you know these, it does not follow that you immediately possess a *habitus* or skill to separate the certain from the uncertain; that you are sharp in your judgement; have the patience to consider the case minutely and acutely; that you are cunning in finding hidden truths, etcetera. Therefore, this skill should be acquired though a continuous training and practice of the rules provided by theories of reason.[60]

The anonymous dialogue *Conversation between a man and a woman about the usefulness of geometry for young students* (1743), likely written by Celsius, explicitly discussed the role of mathematics in the education of a boy.[61] The protagonist of the short text visits a friend during a voyage, and has the opportunity to discuss the usefulness of mathematical education with the friend's wife. The friend's son is soon to leave for university, and the responsibility for his education is soon to be transferred from his mother and father to a tutor. By presenting the wife's position in relation to her son's education, it presents mathematical education as not just an all-male affair. At least as characters in fictional narratives, female subjects of the early modern Swedish

---

[59] "ställa sina tankar efter en förnuftig ordning"; "*Studerande* blefwe från barndomen öfwade at utwidga sina förstånds krafter"; "intet antaga något utan grund och *raison*, samt finna en smak och nöje i *demonstrationer*."; "en färdighet i förståndet"; Celsius: *Arithmetica*, preface. Similar arguments of mathematics as an exercise of reason can also be found in Christian Gustafsson Roman: *Wälmente tanckar om barna upfostring. I anseende till deras studier och lefnad* (Västerås, 1743), 43; Euklides and Strömer (tran.): *Euclidis Elementa eller Grundeliga inledning til geometrien*, till läsaren. For a discussion of mathematics in eighteenth-century Sweden as a realm of certainty, see Dunér: *Tankemaskinen*, 45.

[60] "en rätt *Logica* eller Förnufts-Lära föreskrifwer wäl reglor och *maximes*; men oansedt man dem har sig bekanta, så följer dock intet deraf, at man strax äger en *habitus* eller färdighet, at skilja det wissa från det owissa, det tydeliga från det otydeliga: at man är skarpsinnig at fälla sitt omdöme: har tålamod at saken noga och diupsinnigt öfwerwäga: är slug at upfinna fördolda sanningar, och så widare. Bör altså denna färdigheten förskaffas af en stadig öfning och *practicerande* af de reglor, som FörnuftsLäran gifwer wid handen"; Celsius: *Arithmetica*, preface.

[61] Gunnar Brobergs identifies the author as Anders Celsius in "Then gifte Philosophen eller en man som blyges att wara gift", in *Annales Academiæ regiæ scientiarum Upsaliensis:s. Kongl. Vetenskapssamhällets i Uppsala Årsbok* 26 (1985–86), 83. The similar arguments about mathematical exercises found in the dialogue and in Celsius' *Arithmetica* support this assumption. For a discussion about the role of anonymity in this dialogue, see Elisabeth Mansén: "Samtal emellan en Herre och en Fru år 1743" in Emma Hagström Molin and Andreas Hellerstedt (eds): *Lärda samtal. En festskrift till Erland Sellberg* (Lund, 2014, 158–161).

state related to mathematical education.[62] When describing mathematical studies to the woman, the protagonist urged her to imagine how dancing was learnt:

> I cannot deny, that logics shows us some manners, by which one should think and argue. But thereby one is not given the actual skill to use these rules. Imagine, my lady, that the dancing master, who visits your little daughter, would sit down next to her when he comes, and start to explain in detail how a minuet should be danced: how many steps one should take forwards and backwards, how the figure should be made into a Z, and in addition to this description present everything accurately in a drawing on paper. Do you think, my lady, that after your daughter had learnt all that he had said by heart, and had carefully studied the figure, that she would be able to rise and dance a minuet?[63]

When the woman did not think so, the protagonist pointed out the similarities between mathematics and a bodily exercise such as dancing:

> Your son may learn from logics about all the devices, which one should use to reason; but nonetheless, he would at many occasions never be able to remember these rules. Instead, most of the time he would make false conclusions, take the wrong position, and confusedly relate one thing with another. Therefore, your son should practise making the right conclusions, and diligently train the ways of thinking that logic prescribes.[64]

---

[62] The role of the woman in this dialogue, as a voice of reason in relation to her man, is similar to the role of women in a number of similar mathematical dialogues written by Polhem. See for example: Christopher Polhem: *Samtal emellan en swär-moder och son-hustru, om allehanda hus-hålds förrättningar* (Stockholm, 1745); Christopher Polhem: "Samtahl emällan fröken Theoria och byggmästar Practicus om sitt förehafvande", in Henrik Sandblad (ed): *Christopher Polhems efterlämnade skrifter* vol. 1 (Uppsala & Stockholm, 1947 [undated]), 277–307; Christopher Polhem: "Samtahl och discurs emellan theoria och praxis om mechaniska och physicalska saker huar igenom ungdomen, som der till har lust, kan lära något", in Liljencrantz (ed): *Polhems efterlämnade skrifter* vol. 3, 427–45. In an autobiographical narrative found in a letter, the Swedish female poet Hedwig Charlotta Nordenflycht discussed her passage into womanhood, by her engagement with a mechanicus, in much similar ways to the narratives discussed here. Hedvig Charlotta Nordenflycht: "Autobiographical letter to H:r A.A. von Stiernman, 1745-08-17", *Skrifter* (Stockholm, 1996).

[63] "Jag kan intet neka, at ju Logican visar oss vissa maner, hvarefter man bör tänka och argumentera; men derigenom får man ändå intet en färdighet at bruka dessa reglor i sielfva värket. Inbilla Er, min Fru, at Dansmästaren, som går til er lilla Dotter, skulle, när han kommer, sätta sig brede vid henne, och begynna närligt på at tala om, huru en menuet bör dansas: huru många steg man bör gå fram och tilbakars; huru figuren bör giöras som et Z; ja, förutan denna beskrifningen föreställa altsammans helt accurat uti en ritning på papperet. Tror ni väl, Min Fru, sedan Er Dotter, väl lärdt utan til alt det han sagt, och noga beskådat figuren, at hon då skulle kunna stiga op och dansa en menuet."; Anonymous: *Samtal emellan en herre och en fru, om geometriens nytta för unga studerande* (Stockholm, 1743), 11.

[64] "kan Er Son af Logican få lära at tala om alla de grep, som man bör bruka at raisonera; men lika fult torde han vid monga tilfällen aldrig komma ihog dessa reglor; utan giöra som oftast falska slutsatser: taga saken på galen fot, och invekla det ena med det andra i största confusion. Derföre bör Er Son väl öfva sig at sluta rätt, och flitigt practicera de tankesätt, som Logican föreskrifver"; ibid., 12.

Like Brandt, the protagonist of this dialogue saw mathematics as an exercise, which moulded the mathematician into a man with certain habits. He contrasted this mathematical exercise to Aristotelian logic, another realm of certainty, but one that he considered to be just a set of rules that would not shape the mind of the student. By studying mathematics the son would: "pick up a habit, to make reasonable conclusions about everything." The woman would "then, with pleasure, see him return from academia like a new human being."[65] His choice of likening mathematics to dancing was hardly gratuitous: in the Swedish universities of the time dancing masters regularly taught young noblemen the correct bodily posture and movements that would identify them as belonging to the higher strata of society.[66]

Other writers considered mathematics to be linked to imagination rather than reason. Following classical traditions, in his *Necessary exercises of reason* Rydelius delineated four faculties of the mind [sinnesgåvor]: "the first is reason, the other imagination [...], the third memory and the fourth the will."[67] Imagination was not only "a mistress of all crafts and arts, that are executed with the limbs of the body", it also had much to say in the scholarly sciences, and especially so in mathematics.[68] Rydelius made a distinction between two forms of imagination: "*imaginatio lucens*", enlightened imagination, and "*imaginatio ardens*", burning imagination. *Imaginatio lucens*, Rydelius argued, was "cold, and [...] thus remains in all its operations within the realm of the brain without ever having contact with the heart." *Imaginatio ardens*, on the other hand, played "with such images, which by their nature touches the heart".[69] To clarify his point, Rydelius provided an example that also showed how the mechanicus was to relate to emotions in his work:

> For *example*: When a *mechanicus* pictures, with his vivid *imagination*, all the parts that are necessary in a *machine*, [...] – their shape, order and connections, and the movement, which their construction, and other natural circumstances must effect – such *imagines* can never by themselves awaken any emotions. But if he thinks of the glory or profit, which can be gained

---

[65] "en habitude, at sluta förnuftigt om all ting."; "med nöje se honom komma hem ifrån Academien såsom en ny menniskia."; ibid., 13.
[66] Gustaf Moberg: *Från exercitier till modern idrott* (Uppsala, 1950), 345–46; Dahl: *Svensk ingenjörskonst under stormaktstiden*, 129–30.
[67] "den första är förnuftet, den andra Inbilnings-Gåfwan [...], den tredie Minnet, och den fierde Willian."; Rydelius: *Nödige förnufftz öfningar* vol. 1, § 8.
[68] "en Mästarinna för alla Handwärk och Konster, som drifwas med kroppsens lemmar,"; ibid., preface.
[69] "kalsinnig, och [...] därför stanna med hela sitt spel innom hiärna-kretzen, utan at hafwa mäd hiärtat något at beställa."; "mäd sådana bilder, som af sin natur äro hiärtrörande", Andreas Rydelius: *Nödige förnufftz öfningar* vol. 4 (Linköping, 1720), 42.

Figure 1. The frontispiece of Anders Celsius' *Arithmetica*. The engraving was signed as invented and made [*inven. et fecit*] by J[an] Klopper, the teacher of drawing at Uppsala, and sculpted [*Sc.*] by Carl Meurman, auscultator in the Bureau of Mines. (Photo: Uppsala University Library)

from [the machine], or at least the positive improvement of his knowledge, then the object of his mind is not just a *verum*, but a *bonum*, which directly corresponds to the heart.[70]

In contrast to the mechanicus, Rydelius pointed out how the poet used *imaginatio ardens* when picturing "love, enmity [...], sickness, pain and anguish, or even death itself". Such objects of the mind communicated directly with the heart. According to Rydelius, the same man seldom possessed both cerebral and heartfelt imagination: "generally, the more profound and inventive a *mathematicus*, the slower a *poet*. The more artful a *poet*, the fewer suggestions in *mathesis*". A man's imagination also changed as he matured. Whereas *imaginatio ardens* was stronger among the young, as he entered manhood *imaginatio lucens* gained the upper hand.[71] In a mature man, imagination was thus not governed by basic impulses, but by reason, and mathematics was a means of fostering such an enlightened imagination.

Looking past which parts of a boy's mind were trained by mathematical practice – reason or imagination – all these authors saw mathematics as a technique of discernment that allowed a man to see and reconcile opposing positions. In the Swedish translation of Julius Bernhard von Rohr's work on how to live as a public servant (1728) Nils Rosén (1706–73), who had studied mathematics with Celsius in Uppsala in 1727, summarised this view. He defined reason as the ability "to see the connections between all sorts of truths", and argued that nothing trained it as well as pure mathematics. A mathematically trained man was unmatched in connecting separated truths "because he is a quicker thinker, knows better how to connect reasons, without jumping to conclusions, and is nonetheless of a greater penetration".[72]

The frontispiece of Celsius' *Arithmetica* similarly presented mathematics as a technique of discernment; there, mathematics was portrayed as a mother, surrounded by boys depicted as puttos who diligently studied together (see Figure 1). Thus, like in the anonymous dialogue, women were given an allegorical role as nurturing mothers, overseeing the mathematical education

---

[70] "Til *exempel*: När en *mechanicus* föreställer sig, medelst sin starka *imagination*, alla de delar, som äro nödige til en *machine*, [...] deras skapnad, ordning, sammanhang, och dän rörelse, som af deras sammanfogning, och fler naturliga omständigheter förorsakas måste, kunna dässe *imagines* af sig siälfwa aldrig hos honom uppwäckia någon sinnes rörelse. Men om han täncker på ära häller winst, som där igenom kunna förwärfwas, häller åtminstone på sin wettenskaps lyckeliga förkofring, så blifwer då intet längre ett blott *verum*, utan *bonum*, hans fägnesamma *objectum*, som straxt *corresponderar* mäd hiärtat."; ibid., 42–3.

[71] "Älskog, owänskap [...] siukdomb, pijna och ångest, ja sjålfwa döden"; ibid., 43. "Gemenligen ju diupsinnigare och nyhittigare Mathematicus, ju trögare *Poet*. Ju förslagnare *Poet*, ju fattigare på förslag i *mathesi*."; ibid., 44. For a through, albeit internalist, discussion of imaginatio ardens and lucens in Rydelius, see Wermlund: *Sensus internus och sensus intimus*, 19–24.

[72] "see in i allehanda sanningars sammanhang"; Julius Bernhard von Rohr: *Inledning til klokheten at lefwa, eller Underrättelse om, huru en menniskia, genom en förnuftig sitt lefwernes inrättning, kan bli timmeligen lycksalig* (Stockholm, 1728), 37–8. "ty han är i eftertänckiande hurtigare, wet bättre at sammanbinda skiälen, utan at giöra något språng, och är jämwäl af större penetration"; ibid., 41–2.

of boys. A banner lay wrapped around her legs with the words *"oculis retectis"* (i.e., "with uncovered eyes"). To the left was a picture frame filled with mathematical symbols and to the right, in the background, was Gustavianum, at the time the university building in Uppsala where Celsius was professor. In the lap of mathematics lay a boy, one eye covered by a piece of cloth. Mathematics caringly removed the cloth from his eye; soon he would be able to join the other boys who by the feet of mathematics read a book (perhaps Celsius' own *Arithmetica*). The symbolism was clear: mathematics was a caring mother, who raised boys from youthful blindness into perceptive manhood. In his preface, Celsius explicitly described this process: "Now that *mathesis*, as it were, has removed the inherent cover over the eyes of our reason, then one can see the truth more easily with uncovered eyes."[73] With the inherent cover removed, the boy would mature into a man with a penetrating gaze, and who could connect separated truths of the divine order.

A path to concord or confused indolence?

In his *Arithmetica* Celsius proposed that, because mathematics was "arranged in such a solid order," its diligent study produced a well-ordered mind in the student; it would "dampen" the inclination that "the young generally have for such things, which only touch and please their external senses". And just as mathematics would make the student's mind well ordered, mathematics also made him behave orderly towards his peers. If boys studied mathematics at a young age, "less controversy would arise among scholars; fewer suggestions would be done against the truth, [and] one would hear fewer men who based their arguments on authority, without themselves considering whether it is true or false". In the preface of the *Arithmetica*, and in many other similar prefaces from the time period, mathematics was not just imagined to be related to making knowledge, but also to be a technique of consent and a basis for a harmonious community of knowledge makers. By studying mathematics, a boy would thus learn to see the world in a new way. Such studies would discipline his reason and imagination, so that he would be governed neither by external appearances nor by his own youthful vanity or pursuit of glory. Also, he would learn to see connections between disparate truths, and accept true statements by others without entering into unnecessary conflict. Ideally, mathematics was not only a way to see nature in a new light: it was also a way to relate differently to one's fellow men. The student of mathematics would not only become reasonable or imaginative,

---

[73] "När nu således *Mathesis* hade lika såsom borttagit det medfödda täckelset för wåra förstånds ögon, så kunde man sedan wara beqwämligare at med uptäkta ögon se sanningen"; Celsius: *Arithmetica*, unpaginated preface.

he would also acquire the virtue of concord.[74] As discussed earlier, mathematics and mechanics were Rydelius' chief examples of exercises that fostered a manly enlightened imagination, aligned with the rational theocratic order. He defined mathematical and mechanical creativity as requiring a certain kind of imaginative man, whose creativity was rational rather than emotional. By studying mathematics, a boy could learn to see past glory and private profit. Consequently, he could also become a pious subject, submitted not only to his own reason, but also to his sovereign and ultimately to the divine Lord.[75]

The *Conversation between a man and a woman* discussed mathematics as an exercise of concord in more detail. When the woman of the narrative admitted that she was "afraid [...] that my Son will become stubborn from [mathematical studies], and [that he] will want to do everything as he himself pleases when he thinks he has reasons for it", the protagonist calmed her: "the more he reads Euclid, the more he will understand to respect his reasonable mother; and the more he will accept her rational views."[76] Mathematics was an exercise of reason, and therefore also a means for a boy to discipline his own mind to submit to his reasonable superiors as well as to acquire authority to discipline others. By exercising his reason and his imagination, a boy would become a man who acted in the interest of the state and for the glory of God. But not all mathematics was an exercise of discernment and concord. Even if a boy started his mathematical studies at the appropriate young age, he also needed to study mathematics correctly. While the dialogue presented mathematics as a technique for virtuous performances, it also presented its negative side. The woman of the dialogue presented the two sides of mathematics:

> I wish to have you as arbiter between my husband and me. We are always
> arguing over our son's tutor. I believe that he is a good-natured man, but

---

[74] "en så grundelig ordning förestält"; "dämpas"; "de unga gemenligen hafwa til sådana saker, som allenast röra och beweka deras utwärtes sinnen"; "intet så många stridigheter ibland de Lärda skulle upkomma: intet så många ogrundade inkast giöras emot sanningen: intet höras så många beropa sig allenast på *auctoritet*, utan at tänka sielfwa efter, om det är sant eller osant"; ibid., preface. Compare to Mört's argument discussed in the previous chapter, in Weidler and Mört (tran.): *En klar och tydelig genstig*, preface. Similar arguments for mathematics as an exercise of virtue can also be found in publications by Anders Gabriel Duhre. For a discussion of these, see Chapter 5. These arguments can also be found in the works of early modern philosophers, such as for example René Descartes, Blaise Pascal and Gottfried Leibniz, that considered mathematics and natural philosophy to be "powerful tools for living a good and virtuous life"; Jones: *The good life in the scientific revolution*, 1–3, quote from page 1. Sverker Lundin has also discussed Celsius's view of mathematics as an exercise of reason and virtue in Lundin: *Skolans matematik*, 155–7.

[75] More than twenty years later, in 1746, Eric Eklund referencing Rydelius echoed these views, and argued that tutors in mathematics should take note of boy with a strong imagination," Eklund: *Upfostrings-läran*, 251.

[76] "rädder, sade Frun, at min Son blir härigenom envis, och vil giöra alt efer sit hufvud, när han tycker sig hafva raisoner dertil"; "ju mera han läser Euclides, ju mera förstår han at vörda en så raisonable Mor; och ju mera skal han finna sig uti hennes förnuftiga förestälningar."; Anonymous: *Samtal emellan en herre och en fru*, 14.

my husband says that he informs our child of a bunch of unnecessary things. As our son is meant to become a priest, he says, it is futile to torment him with triangles and circles and other fads and fancies, which [my husband] calls novelties [nyheter].[77]

The protagonist first agreed: "Your husband is right that triangles and circles hardly will help your son eventually to deliver a sermon or to sing mass," but he continued, "Geometry should be studied because it provides the habit of reasoning clearly."[78] The woman accepted this argument, but she explained that her husband was of the opposite opinion. He believed that:

> such figures and lines, if anything, make people confused, and [he] exemplifies this by a certain *mathematicus*, whom I also thought was somewhat strange. However, I do not believe that mathematics was the cause of this. Instead, [this man] had stayed too much indoors by himself, reflecting upon his figures, which had made him unaccustomed to the company of others.[79]

The dialogue expressed two opposing opinions of mathematics as an educational practice. On the one hand, the protagonist gave voice to the belief that mathematics could raise a boy into a diligent man, aligned with a rational social and natural order. On the other hand, the husband argued that not only was mathematics of no use to most boys, it might even make them into confused men who failed to meet the demands of everyday life. What was at issue in these texts was the relationship between mathematics, the virtue of diligence and the epistemic virtues of *theoria* and *praxis*. As seen earlier, many authors presented mathematics using the metaphor of bodily exercise and they placed mathematics in the realm of *praxis*, as opposed to logic that was pure *theoria*; they all imagined mathematics as hard work, rather than a simple mind game. In a Lutheran state, where diligence was a manly virtue, this way of discussing mathematics may not be that surprising: to discuss mathematics in relation to *praxis* was yet another way to present it as useful work that would foster a virtuous man. The dialogue thus sought to convince the reader that mathematics encompassed *theoria* as well as *praxis*, and that it was an exercise not only of concord but also of diligence.

---

[77] "Min Herre! sade hon, jag vil taga Er til Domare emellan mig och min man. Vi ha alltid en dispute om Præceptoren för vår gosse. Jag tycker, det är en beskedlig karl nog; men min Man säger, at han informerar vårt barn i en hop med onödiga saker. Efter vi ämnat vår son at bli prest, så säger han, at det är fåfängt, at plåga honom med trianglar och cirklar, samt annat grillerväsende, som han kallar nyheter"; ibid., 3.

[78] "tycker Er man rätt nog, at trianglar och circlar litet hielpa Er son med tiden at giöra en predikan och siunga en Mässa"; "men derföre bör Geometrien läsas, at man kan bli van at raisonera redigt"; ibid., 4.

[79] "sådana figurer och linier giöra folket snarare förvirrade, och drar til exempel en viss Mathematicus, som jag tyckte ock var något underlig; men jag menar, at intet mathematiquen varit dertil orsaken, utan at han suttit för mycket inne i enslighet och speculerat på sina figurer; hvarutaf han blifvit ovan at umgås med annat folk."; ibid.

For many authors, mathematics was dangerously close to the realm of *theoria* and to speculation far removed from everyday application.[80] If unchecked, a boy who only studied pure mathematics, and who studied the wrong kind of mathematics, would become anything but a mature manly subject. Rosén's translation of Rohr warned that a reasonable and practical man should not "focus for too long on the numbers or the lines." Instead, he should "acquire from mathematics only its accurate way of learning, which you can later can, and should, apply on other sciences [*wetenskaper*]". One should not attempt to become too proficient in pure mathematics, as "a *purus putus mathematicus* is often unskilled in the functions of everyday life."[81] Similarly, Duhre gave voice to critics who questioned whether it was possible to combine serious mathematical studies with other exercises, because mathematicians are "unskilled in everything which does not comply with their theorems", and because mathematics "is so extensive that anyone who wants to make considerable progress, will need a whole lifetime".[82]

In Aristotelian science, *theoria* was a set of certain systematic knowledge, in contrast to *praxis* that refused to be defined by a solid deductive system. But in seventeenth- and eighteenth-century discussions of *theoria* and *praxis* the role of the two grew more complex. By the seventeenth century, attacks on Aristotelian epistemology had overturned the relationships between these terms. The new meaning of this dichotomy, established by the anti-scholastic humanist rhetoric of the Renaissance, can be found in the educational reformer Petrus Ramus. For Ramus, education should focus on *praxis* and *usus* (that is, use) rather than *theoria* and *doctrina* (rules and precepts). *Praxis* was related to the natural world and by consequence productive.[83] At the same time, *theoria* was a guarantor of stability: the rules and precepts of theological and philosophical systems were means of integrating the natural and social order. Thus, unsurprisingly, advocates of useful knowledge – knowledge that fostered productive and good subjects – generally imagined such knowledge to integrate these two epistemic virtues. For example,

---

[80] This scepticism of pure mathematical theory was not new to the eighteenth century. Ingemarsdotter has pointed out that Ramus was "sceptical of the utility of pure mathematical theory"; Jenny Ingemarsdotter: *Ramism, rhetoric and reform. An intellectual biography of Johan Skytte (1577–1645)* (Uppsala, 2011), 35. Similarly, Gaukroger has discussed how in the sixteenth and early seventeenth centuries the usefulness of mathematics was a disputed question, Gaukroger: *Francis Bacon and the transformation of early-modern philosophy*, 20–7.

[81] "uppehåll tig icke alt förlänge med talen och *linierne*"; "utan lär ur *Mathesi* allenast deras *accurata* läre-sätt, som tu sedan i andre wetenskaper bör och kan *applicera*."; Rohr: *Inledning til klokheten at lefwa*, 38. "en *purus putus Mathematicus* är ofta helt oskickelig til menniskliga lefwernets förrättningar."; ibid.

[82] "oskicklighet til alla andra syslor, hwilka intet komma öfwerens med deras satser"; Anders Gabriel Duhre: *Förklaring öfwer des tilförende uthgifne Wälmente tanckar angående huru han tillika med sin broder är sinnad at utan almenna bestas betungande uppå deras egit äfwentyr uprätta ett laboratorium mathematico-oeconomicum* (Stockholm, 1722), 10. "Dess utan är denna wetskap så widlyftig, at den som der uti wil komma något wida, behöfwer der til sin hela lifstid"; ibid., 11.

[83] Smith: *The body of the artisan*, 17–20; Ingemarsdotter: *Ramism, rhetoric and reform*, 32–3.

Rydelius considered *theoria* and *praxis* to be stages in a process of maturity: whereas a boy could learn useful knowledge from reading the theoretical principles found in books, a man should learn such knowledge from the everyday *praxis* of carrying out his office.[84]

Lindqvist has pointed out that the aim of the mathematical and mechanical practitioners of eighteenth-century Sweden was more complex than the "the common historical generalization [...] that they tried to 'unite theory and practice'."[85] However, the unification of *theoria* and *praxis* was a reoccurring theme when mechanical practitioners presented their work to contemporary audiences in the German and Scandinavian states. In order to understand the interrelated roles of these two concepts, we must be sensitive to the meanings that historical actors ascribed to them. To early modern actors, the reoccurring theme of bridging these categories carried an implicit promise of conjuring forth an orderly and pious state. Given the place of mechanics in early modern epistemologies, mechanical practitioners had a number of possibilities to argue that their art was suitable for the education of virtuous and useful subjects. As argued by Bennett, by the sixteenth and seventeenth centuries, mechanics came to be characterised as a type of practical mathematics. Such practical mathematical knowledge (including, besides mechanics, optics, fortification and navigation, for example) bound together pure mathematics (i.e., geometry and arithmetic) and crafts. A long European tradition defined such practical mathematics as involving both mind and hands: they were mediating sciences (*scientiae media*) that bridged knowledge of nature (*physica*) and mathematical knowledge. As a form of practical mathematics, mechanics designated on the one hand "manual" as well as "practical" activities, the design and construction of machines, and on the other a doctrine about the natural world; *mechanica* was both identified in the work of craftsmen and studied in universities as a part of natural philosophy and natural theology.[86]

Because of the intermediary position of mechanics in early modern epistemology, mechanical practitioners could thus convincingly argue that mechanics was a conciliatory exercise that fostered concord in a boy and, at the same time, that it involved hard work that cultivated diligence. In short, they could argue that correctly taught mechanics fostered the full range of virtues of a good manly subject. In his *Anleitung zu nützlichen Wissenschaften* (1700) – a well-known introduction of useful sciences for boys and young men – the

---

[84] Rydelius: *Nödige förnufftz öfningar* vol. 1, § 6.

[85] Lindqvist: *Technology on trial*, 67.

[86] Bennett: "The mechanics' philosophy and the mechanical philosophy"; Jim Bennett: "The mechanical arts", in Katharine Park and Lorraine Daston (eds): *The Cambridge history of science* vol. 3: early modern science (Cambridge, 2008), 163; Ingemarsdotter: *Ramism, rhetoric and reform*, 168. On the Aristotelian roots of the concepts of *scientiae media*, see Walter Roy Laird: "Robert Grosseteste on the subalternate sciences"; *Traditio* 43 (1987), 147; Jeremiah Hackett: "Roger Bacon on scientia experimentalis", in Jeremiah Hackett (ed): *Roger Bacon and the sciences. Commemorative essays* (Leiden, 1997), 286.

German mathematician Ehrenfried Walter von Tschirnhaus discussed this issue explicitly. Tschirnhaus rhetorically proposed *praxis* and *theoria* as two routes to learning useful practical mathematics, but argued that neither of these paths alone was sufficient. Only a *via media*, that "almost always joined *praxis* with *theoria*" would let young boys acquire useful knowledge from a young age.[87] By presenting practical mathematics as a *via media*, Tschirnhaus not only performed a classificatory exercise, he also claimed a social space for mechanics. His classifications served a didactic end: they were means of marking out the route to a certain form of useful manhood.[88] When early moderns imagined a mechanicus, they saw him as treading this *via media*, leading to the life of a useful and pious man. In this vein, Polhem in 1740 argued that three parts "belong to a competent *mechanicus*, namely: first, a natural ability to invent something new; second, an aptitude and inclination for theory; and third, a practical handiness".[89]

Wrongly approached, however, mathematics would neither be a path to concord nor diligence. Instead, it would be an exercise of vice and confusion. Duhre presented some examples of such dishonest mathematicians:

> Consider how lewd fellows are recognised as great *algebraists* by many people, because they have calculated some *examples*, which they have collected from strange arithmetic books, in order to amuse the guests of inns and taverns and thus to earn something with which to buy a stoup of beer.[90]

Duhre considered it just as problematic to practise mathematics in the wrong crowd: "Many lewd comrades are greeted as great *geometricians*, because they have taught simple journeymen how to add some geometric tricks to their work."[91] These were all examples of how, wrongfully execut-

---

[87] "da fast immer *praxis* mit der *Theoria conjungi*ret wird,"; Ehrenfried Walther von Tschirnhausen: *Gründliche Anleitung zu nützlichen Wissenschafften absonderlich zu der Mathesi und Physica. Wie sie anitzo von den Gelehrtesten abgehandelt werden* (Frankfurt & Leipzig, 1700), 15. Tschirnhaus was Rydelius' recommended reading for Swedish students, and he pointed out that his little book was available and easy to buy for a Swedish student. Andreas Rydelius: *Nödige förnufftz öfningar* vol. 3 (Linköping, 1719), 11. For a general discussion of Tschirnhaus' pedagogical method, see C. A. Van Peursen: "E. W. von Tschirnhaus and the ars inveniendi", *Journal of the history of ideas* 54:3 (1993), esp. 398.

[88] Alan Gabbey: "Between ars and philosophia naturalis. Reflections on the historiography of early modern mechanics", in J. V. Field and Frank A. J. L. James (eds): *Renaissance and revolution. Humanists, scholars, craftsmen, and natural philosophers in early modern Europe* (Cambridge, 1993), 137. Stephen Gaukroger has pointed out how such classifications were often made to a didactical end, Gaukroger: *Francis Bacon and the transformation of early-modern philosophy*, 18.

[89] "höra en god *Mechanicus* til, nemligen I:o En gåfwa af naturen at kunna finna på något nytt. II:o Hug och lust til *theorien* och III:o Et snält handalag i *practiquen*"; Polhem: "Tankar om mekaniken", 193.

[90] "för ögonen ställa, huruledes liderliga foglar af många warda ansedda för stora Algebrister, derföre at de uträknat några exempel, som de tagit utur konstiga räken böcker, på det de der med på kiälrar och krogar skulle kunna roa giästerna, och således förtiena något at kiöpa sig et stop öhl före."; Duhre: *Förklaring*, 11.

[91] "warda sådana liderliga sällar af mången helsade för stora Geometræ, derföre de hafwa underwisat enfaldiga handtwärcks gesäller, huru de böra lempa några gemoteriska [sic] handgrep til sina syslor"; ibid.

ed, mathematical studies would fail to teach a boy manly virtues. They also show that mathematics was not necessarily seen as useful knowledge and that the role of the mathematical practitioner was morally ambiguous. As pointed out by Katherine Hill, in the seventeenth century the role of the mathematical practitioner was under negotiation. Was mathematical authority defined by knowledge of arithmetic and geometry, or by the use of instruments and the practical mathematical arts?[92] In the prefaces, there was a similar negotiation: was mathematics a diligent exercise of consent and discernment, or was it just a means of theoretical confusion and false promises? In eighteenth-century mathematical works, a bad mathematical practitioner was identified primarily through two traits. First, he practised mathematics for personal gain and not to reconcile opposing positions and serve the *publicum*. Second, he used mathematics as a collection of useless tricks and examples in order to avoid real work. That is, he failed to encompass the virtues of concord and diligence.

For a boy to become a useful subject from mathematical practice, he should instead perform it in line with these manly virtues. The boundary between good and bad mathematics was drawn between techniques that contemporary audiences perceived as useful and useless. But to identify this boundary was harder than one would guess, and actors drew it differently. Also, even seemingly useless forms of mathematics might be diamonds in the rough. The woman of the anonymous dialogue stated that she had "heard how geometric people ponder things, that never provide any utility."[93] The protagonist replied:

> That may be true, my lady, [...] but that hurts no one. Let some people use their brain to find one truth after another in all sorts of figures. There might come a time, when all these ponderings become very useful. Our great mechanicus Polhem has in our time shown the application of a bunch of curved lines to bridges and machines among other things.[94]

Convinced by this argument, the woman answered that one could maybe liken "these truths, that are still fruitless, to some of our beautiful flowers [...]; perhaps some Doctor Linnaeus may come along and show what they are good for."[95] In this dialogue, this kind of mathematics – whose curved lines were made useful just as beautiful flowers were put to use by an œconomically minded natural historian – carried meanings other than pure

---

[92] Katherine Hill: "'Juglers or schollers?'. Negotiating the role of a mathematical practitioner", *The British journal for the history of science* 31:03 (1998), 253–74.

[93] "Jag har dock hördt, sade Frun, at det här Geometriska folket grubla på saker, som de aldrig veta nogon nytta af."; Anonymous: *Samtal emellan en herre och en fru*, 9.

[94] "Nog är det vist, min Fru! svarade jag; men det skadar intet. Låt man nogon bry sin hierna dermed, at utleta den ena sanningen efter den andra uti allehanda figurer. Det kan väl komma en tid, då dessa griller blifva mycket nyttiga. Vår stora Mechanicus POLHEM har i vår tid nog vist prof af en hop med krokuga liners application til broar och konstbygnader, med mera."; ibid.

[95] "dessa ännu fruktlösa sanningar vid en del af våra skiöna blommor, [...] det kommer väl nogon Doct. LINÆUS, som visar, hvad de duga til."; ibid., 9–10.

mathematical speculation. Similarly to the works by Duhre and Rosén, it thus expressed an anxiety over the socio-epistemic status of mathematics. To pick up an analogy from the dialogue, virtuous mathematics was like dancing in several senses. Not only did it require diligent exercise, but the performance of mathematics also demanded that practitioners constantly mind their steps. Only when performed in the rhythm of the theocratic order would mathematics excel. On the one hand, mathematics was an exercise of reason and concord. On the other hand, it was a treacherous landscape where a boy could get lost among useless curves. Only boys who joined the virtues of concord, diligence, *theoria* and *praxis* in their performance of mathematics would enter useful and virtuous manhood. Authors, such as Celsius and Brandt, argued that such performances were possible because the pure mathematical arts of geometry and arithmetic were more akin to bodily exercise than a rote following of rules. Others maintained that boys who combined the study of geometry and arithmetic with everyday work followed a safer route to manhood. One such safer path was the *via media* of the mechanicus. When older men imagined how a boy was to become a mechanicus by following this route, they envisioned this exercise to consist of the integrated exercise of mathematics and crafts.

Crafts and the manliness of hard work

Pure mathematics always risked being morally suspicious, due to its connotations of speculation. Therefore, the narratives of experienced mechanical practitioners imagined mechanics to embody mathematics in an able body fostered by the exercise of crafts. But similar to how mathematical education was a complicated and morally ambiguous dance, the integration of crafts and mathematics was not a completely straightforward business. Because crafts were primarily associated with craftsmen, mechanical practitioners' reflections on crafts were ripe with anxieties of being misidentified. Interestingly, their concerns were similarly related to the ambiguous nature of mathematics, discussed earlier. How could mechanically inclined boys learn crafts without becoming entangled in subordinate relationships with craftsmen, and without being led on to a life other than that of the mechanicus? As in the case of mathematics, these concerns revolved around the role of the virtues of diligence, concord, *theoria* and *praxis* in the exercise of crafts.

In early modern Scandinavia, crafts were often categorised as *slögd*. The word stemmed from the ancient Swedish word *slöghð* that meant slyness, cunning or artfulness. In eighteenth-century Swedish, the word developed into – on the one hand – *slug*, keeping these connotations of cunning, and – on the other hand – *slögd,* which mainly denoted arts and handicrafts as well as industry.[96] The way *slögd* and *slug* were etymologically linked thus translates well into the English "crafts" and "crafty", respectively. Crafts was not a homogeneous concept. It consisted of diverse identificatory practices of knowledge and material production with different class and gender connotations.[97]

Fletcher has pointed out that in early modern England, men were concerned with "whether their boys would acquire the secure manhood" demanded of a mature man, and boys' physical bodies were considered to be "vulnerable to the pressures of a blurred gender system." By engaging in unsuitable activities, boys could become effeminate, a term referring to "unmanly weakness, softness, delicacy and self-indulgence."[98] Similar concerns can be seen in the Lutheran culture of eighteenth-century Sweden, where hard work was central to the performance of manhood and the acquisition of a manly body.[99] In his *Characters. Or portraits of persons* from 1754, Abraham Sahlstedt spelled out this ideology of work in full. He explained that because "men commonly possess greater strength than women", a man who "shies away from work, who does not stand any toil or discomfort, and who attends to the attractiveness of his body, is called effeminate [*qwinlig*]" and could also be called a "weakling." Sahlstedt pointed out that "he who by nature has a weak body and delicate limbs is excused, if he cannot bear hard work." Such a man could not be called "effeminate," because "he adjusts his behaviour and his work to what his nature endures and what is respectable for his sex." But for those who "possess manly strength," weakness came from bad upbringing. According to Sahlstedt, effeminate men were "the most unfit members of a society."[100] To carry out hard bodily work according to one's ability was essential for a man who wished to participate as a useful and fit member of the state.

---

[96] "Slöjd", *SAOB* <http://g3.spraakdata.gu.se/saob/> [accessed 4 April 2015].

[97] Albert Wiberg discussed these sets of techniques as "manlig slöjd" and "kvinnoslöjd", Albert Wiberg: *Till skolslöjdens förhistoria. Några utvecklingslinjer i svensk arbetspedagogik intill 1877. Tiden intill 1861* vol. 1 (Stockholm, 1939), 7.

[98] Fletcher: *Gender, sex and subordination in England 1500–1800*, 87. Steven Shapin has argued that there existed notions of "manly nobility of manual work" among the men in the Royal Society of early modern England. As pointed out by Shapin, Robert Boyle presented "physical labor as a Protestant antidote to dangerous idleness"; Shapin: *A social history of truth*, 190.

[99] As shown by Kekke Stadin, in the sermons of the clergy state diligence was the most commonly mentioned virtue in relation to men, Stadin: "Att vara god eller att göra sin plikt?", 230. Also, Andreas Marklund has identified "industriousness" as a "fundamental component" in the peasant masculinity in eighteenth-century Sweden, Andreas Marklund: *In the shadows of his house. Masculinity & marriage in Sweden, c. 1760 to the 1830s* (Florence, 2002), 135.

[100] "i allmänhet at tala, det männer äga större styrka, än qwinnor"; Theofrastos and Abraham Magni Sahlstedt: *Caracterer, eller Sede-bilder af människor* (Stockholm, 1754), 85. "en man, som undandrager sig arbete: som icke tål wid någon möda eller obeqwämlighet, och är sorgfällig om sin kropps behaglighet, blifwer kallad qwinlig"; "en Wekling."; "Den som af naturen fått en swag kropp och späda lemmar, är at ursäkta, om han ej kan uthärda något drygt arbete."; "qwinlig,"; "han tillika lämpar sit upförande och sina göromål efter hwad dess natur kan tåla, och dess kön anstår."; "äger manlig styrka,"; "de odugligaste människor i et Samhälle."; ibid., 85–7.

70

Crafts were considered educational for children of both the noble and the poor, and for women as well as for men. Still, these groups should not engage with crafts in the same way. Different persons – of separate strata of society, or of different gender – should work differently. The question was not whether or not a person should perform bodily work, but *what* exercises were fit for him or her. Manual work was an expression of the earlier discussed virtues of a good subject. Consequently, to craft objects was to form oneself in relation to one's contemporaries. By the manner in which they engaged with manual work, the subjects in eighteenth-century Sweden expressed their social belonging: as poor or rich; childish or mature; manly, feminine or effeminate. In the earlier quotes from Sahlstedt, we see how hard work was linked to manliness. For boys, hard work was a manly performance of diligence that would eventually distinguish them from women and effeminate men.

An exercise of diligence

The commemorative speeches of the Royal Academy of Sciences repeatedly discussed crafts and bodily exercise. There, bodily exercise was not only seen as a way to strengthen the body: it was also expressed as a means to make the mind diligent. Carl Fredrik Mennander described how Gabriel Lauræus, later Professor of Theology at Åbo, practised mechanical arts as a boy. At a young age, he had "with pleasure attentively visited the workshops. There, he sharpened his reason, and at home he imitated the techniques that he had seen being used." Soon, Mennander continued, "he was an artisan in more than one [art]; not a common one but one who could improve the arts through ingenious and useful inventions."[101] In another commemorative speech at the Swedish Academy of Sciences, in 1768, Strömer related a story of how Samuel Enander and Samuel Klingenstierna, when children, used "to carry bricks in a wheelbarrow: that they eventually increased the load, and finally became so accomplished, that the roughest farmhand could not nudge the barrow, which they easily pulled." This activity was not only taken as a tale of how Klingenstierna attained a manly strength at a young age, but also as "an evidence of endurance." Like Lauræus, Strömer imagined how such exercises that "strengthened the body" did not only "contribute to good health": they could "also improve

---

[101] "besökte verkstäderna med nöje och upmärksamhet. Han skärpte der sit förstånd, och efterapade hemma de handalag han sett brukas."; "Innan kårt var han i flera än en konst mästare, icke någon vanlig, utan sådan, som kunde förbättra konsterna genom sinrika och nyttiga upfinnelser."; Carl Fredrik Mennander: *Åminnelse-tal öfver theol. doct. dom-probsten och theol. professor primarius vid kongl. academien i Åbo, herr Gabriel Lauræus, på kongl. vetenskaps academiens vägnar, hållit den 19 december 1755* (Stockholm, 1756), 11.

the rational powers, which always suffer when the body becomes soft and weak through indolence and sedentariness".[102]

As pointed out earlier, in the coming-of-age narratives of mechanics, bodily exercises in general and crafts in particular were instrumental in making a boy into a mechanicus. For Polhem, crafts with masculine connotations, such as "carpentry, turning, [and] crafting", were better bodily exercises for a boy than games such as "playing, hitting balls, [and] throwing bowls". Although games could improve the circulation of the blood, they had no further utility. Crafts, on the other hand, were bodily exercises that exercised the body while also fostering a useful subject.[103] The exercise of crafts not only established a good working relationship between mind and body, but also aligned a boy with the political order. Similar to how the authors in the previous section considered mathematics to be useful, because it was a means of making a virtuous practitioner, Polhem thus considered crafts to foster a boy into a pious and useful man. In a boy's exercise of crafts, the regime of the individual body harmonised with the demands of political order.[104]

Interestingly, when mechanical and mathematical practitioners argued for the usefulness of crafts, they gave them connotations similar to those given to pure mathematics by, for example, Brandt and Celsius: they were diligent exercises that improved the practitioner. As such, they formed the mind and body of a boy to correspond to the ideals of useful manhood. Still, crafts were not necessarily a virtuous exercise: like mathematics, they also had connotations of rote repetition and uselessness. If a boy were to learn how to perform crafts like a useful man, these older men imagined, he needed to distinguish himself from the common craftsmen.

How to learn crafts without becoming a craftsman

Men were not a homogeneous group, and men as a whole were hardly supposed to engage with hard work in the same way. In works on mathematical and mechanical didactics, negotiations on how a mechanically apt boy

---

[102] "föra tegelstenar uti en Skott-kärra: at de småningom ökat lasten, samt sluteligen kommit så vida, at den grofvaste bonde-dräng ej kunde få den lastade kärran af stället, som de med lätthet drogo."; "Et prof af ståndaktighet"; "som stärkte kroppen"; "bidrager til en god hälsa"; "utan äfven hjälper förstånds-krafterna, som altid lida, när kroppen genom maklighet och stilla sittande blifver veklig och svag."; Mårten Strömer: *Åminnelse-tal öfver kongl. maj:ts troman, stats-secreteraren … Samuel Klingenstjerna, på k. vetensk. academiens vägnar hållit, den 27 jul. 1768* (Stockholm, 1768), 6.

[103] "snicka [sic], swarfwa, slögda"; "spela, slå boll, kasta klot"; Polhem: "Tankar om mekaniken", 195.

[104] Andrew Warwick has pointed how, over the first half of the nineteenth century, students of "mixed mathematics" started to integrate bodily exercices such as taking walks into their daily routine as a technique to cope with a high academic work load, see Warwick: "Exercising the student body"; Andrew Warwick: *Masters of theory. Cambridge and the rise of mathematical physics* (Chicago, 2003), 191–200. It seems as if bodily exercise, in the form of crafts, held similar connotations of diligence and order among Swedish early eighteenth-century pratical mathematicians.

should relate to bodily work – and especially crafts – were a reoccurring theme. In his textbook on arithmetic, Brandt pointed out how many viewed mechanics "as merely an art of craftsmen".[105] To engage with crafts risked reinforcing these views and such exercises also risked making a boy a craftsman. But, as pointed out by Duhre, young students of mechanics and mathematics should "not learn crafts to become craftsmen".[106] With this seemingly paradoxical statement, Duhre more or less summarised the role of crafts in the education of a mechanically apt boy. According to these authors, the mechanicus was related, but still distinct, from mere craftsmen.

When learning crafts, boys who aimed to become mechanici were not expected to only practise them alone but also to visit workshops of crafts-men in order to learn the more difficult techniques. Such techniques could be "planing [or] attaching parts by a right angle using a ruler or a set square," techniques that could only be acquired by interacting with crafts-men in their own environment. In the craftsman's workshop, young stu-dents of mechanics could learn how to use these "theoretical instruments" of the craftsmen, whose uses were difficult to figure out on one's own.[107] But to learn crafts in craftsmen's workshops was not a straightforward af-fair. In order to learn a craft from a master craftsman, a boy commonly needed to become his apprentice and thus enter into a position subordinat-ed to him. To enter into such a relationship with a craftsman required of a young boy a dedication to make himself in the craftsman's image. There-fore, authors such as Polhem and Duhre imagined that it would be impossi-ble to study crafts in order to become something other than a craftsman.[108] While they expected a mechanicus to be superior to a craftsman, they con-sidered it impossible to foster and maintain such a superior role in spaces controlled by master craftsmen. To Polhem, such a relationship would be unacceptable for craftsmen and mechanically apt boys alike:

> No carpenter or master builder, who has been trained by craftsmen, can
> stand to have any disciples subordinated to him other than those similar to
> how he himself once was. Such apprentices [of craftsmen] are not always

---

[105] "för en blott handtwärkare konst"; Brandt: *Mathesin universalem*, preface.

[106] "icke böra lära handtwärcken til den endan, at de skola blifwa handtwärckare"; Anders Gabriel Duhre: *Wälmenta tanckar, angående huru han tillika med min broder är sinnad, at utan almenna bestas betungande, uppå wårt egit äfwentyr uprätta et laboratorium mathematico-oeconomicum* (Stockholm, 1722), 5.

[107] "hyfla och foga rät efter *linial* och winkel,"; "*theoreti*ska *instrumenter*"; Polhem: "Tankar om mekaniken", 195. See also Albert Wiberg: *Christopher Polhem som slöjdpedagog* (Gothenburg, 1938).

[108] The dilemma of mechanici of whether or not to enter such crafts spaces, can be seen as simi-lar to how Andrian Johns describe how "the gentleman who entered the world of the Stationers was reducing himself to just one participant in a collective of crafts operatives", and how "be-coming an author meant losing one's self"; Adrian Johns: "The ambivalence of authorship in early modern natural philosophy", in Mario Biagioli and Peter Galison (eds): *Scientific authorship. Credit and intellectual property in science* (New York, 2003), 79. In Polhem's case of young mechanical practitioners, however, what was at stake was not so much that they had a self to lose, as the fact that integration into a collective of craftsmen would set them on a way of life than would make them something else than a mechanicus.

of the quickest kind [...], whereas [mechanics] requires the quickest students there is.[109]

A mechanically apt boy should paradoxically keep a distance from craftsmen while engaging with their techniques. Such a boy needed to avoid entering into a subordinate relationship with a master craftsman and therefore he should also avoid practising certain crafts. While the hard work of craftsmen was manly, Polhem also considered it to be signifying of a lack or reason:

> The method that craftsmen use to teach the young is suitable only for thick-headed idiots, who want to be treated like slaves until a lengthy training rather than reason has made them skilled in making a thing similar to others, without further thought of improvement.[110]

Although a boy should be exposed to such hard work, it was to be governed by reason if it were to shape him into truly useful manhood. As discussed earlier, mathematics was one way of gaining such a power of reason. In a treatise on watermills, Elvius imagined that his mathematical method set him apart from craftsmen; mathematics allowed him to discern the unique designs that were needed in a certain context, and to invent the exact construction needed in a specific situation.[111] Similarly, for Polhem, although "a craftsman and a *Mechanicus* do similar work; it is nevertheless made with a noticeable difference in time, quality and cost". Because of his mathematically trained reason, a mechanicus was a captain of craftsmen: "Similar to how a crowd of Soldiers can accomplish little with their manliness without a sensible captain, the whole lot of craftsmen cannot make anything extraordinary without a good *mechanicus*."[112] In the words of Duhre, it was: "through the *application* of *mathematics* to the acquired skill that the light of day [could] be brought forth, to the incredible utility of human kind."[113]

---

[109] "ingen hantvärkare eller byggmästare som lärt på hantvärkarevijs, tåhl andra läriungar under sig än han sielf varit och sådana icke altijd ähro af det quikaste slaget [...] der likväll dena vettskapen fodrar dee aldra quikaste subjecta som fins"; Christopher Polhem: "Berättelse om Fahlu grufvas tillstånd", in Henrik Sandblad (ed): *Polhems efterlämnade skrifter* vol. 1, 39–40.

[110] "Den method som hantvärkrar bruka att lära ungdommen, tiänar allenast för tiocka dumhufvun som vill på trähldoms vijs vara hantterade till dess en långlig öfning mehr än förståndet giort dem skickeliga att förrätta en ting lijka, utan vijdare efftertanka till förbättring."; Christopher Polhem: "Mechanica practica eller fundamental byggarekonst", in Sandblad (ed): *Christopher Polhems efterlämnade skrifter* vol. 1, 83.

[111] Pehr Elvius: *Mathematisk tractat, om effecter af vatn-drifter, efter brukliga vatn-värks art och lag* (Stockholm, 1742), 109. Compare to how in early modern England, mathematical practitioner defined himself in relation to craftsmen, as discussed by Stephen Johnston: "Mathematical practitioners and instruments in Elizabethan England", *Annals of science* 48:4 (1991), 325.

[112] "en Handtwärkare och *Mechanicus* hafwa lika syslor; men det sker likwist med en så märkelig skilnad i tid, godhet och omkostning"; "Likasom en hop med Soldater kan föga ting med sin manlighet åstadkomma, utan en förståndig höfwitsman, så kan ock hela hantvärkshopen icke idka något synnerligit, utan en god *mechanicus*"; Polhem: "Tankar om mekaniken", 193.

[113] "igenom *Mathematiquens Application* bifogad til förfarenheten kunna dagsliuset framdragas til menniskliga slächtets ännu otroliga gangn."; ibid., 13–14.

Also, in order to distinguish himself from mere craftsmen, a mechanically apt boy should shun heavy crafts, to which reason was difficult to apply, such as "woodcutting and other crude woodworks", and instead master finer carpentry techniques needed to build models. Interestingly, although hard work was a key part of the performance of manhood, in order to become the superior of craftsmen the mechanicus should not work *too* hard. Making such a discriminatory choice, between heavier and finer techniques, would be impossible for an apprentice, argued Polhem. Therefore, the education given by the guilds was not viable for a young man wishing to learn mechanics.[114] In one of his autobiographical manuscripts, Polhem imagined how he himself had been marked by a complicated relationship to hard work. He reminisced that in 1688, when he was 27 years old, he had proved himself a mechanicus by repairing the astronomical clock of the Uppsala cathedral, a mechanical machine placed in the symbolic centre of the Swedish Lutheran theocratic order. When performing this work, "nothing was more annoying, than the fact that I received little help with the heavy work in the church". Instead, he had carried "all heavy pieces and rods" himself. When seeing his heavy work, the students in Uppsala had identified him, not as one of their peers but as "the Professor's blacksmith". Their opinions, Polhem argued, had even forced him to change lodgings, to a location further from the university.[115] In his autobiography, this narrative of hard work became not only a way for Polhem to reflect on the anxiety-ridden performances required of a mechanicus, but also a way to present himself as different from both craftsmen and the other students of the university.

And so the texts discussing the education of a young mechanicus went back and forth. On the one hand, the boy mechanicus needed to engage in the practices of craftsmen and to enter their workshops; on the other hand, he should keep his distance from the spaces of craftsmen, and the rote work and hierarchical relationships of the guilds. In a study of the French *Encyclopédie*, William Sewell has made an observation of the authors' relationship to crafts, which he described as "double-edged." On the one hand, the encyclopaedists wished to show how "mechanical arts" were "complex and subtle achievements of human intelligence". On the other hand, they "believed that the mechanical arts as currently practiced – and the artisans who practiced them – were in need of considerable improvement." For the

---

[114] "hugga med bijhla och anat grofft timermansarbete"; Christopher Polhem: "Falu grufvas tillstånd", 39–40, quote from 40. This role of the mechanicus can be seen as a close parallell to the developing identity of the chemist in the mid-eighteenth century. See Hjalmar Fors: "J. G. Wallerius and the laboratory of enlightenment", in Hjalmar Fors, Enrico Baraldi, and Anders Houltz (eds): *Taking place. The spatial contexts of science, technology and business* (Sagamore Beach MA, 2006), 20.

[115] "var ingen ting förtreteligare än det at iag vijd det grofva arbetet i kiörkan fick ringa hielp utan moste sielf bära alla grofva stycken och stenger"; "Professorens Smed"; Christopher Polhem: "Lefvernesbeskrifning", Bengt Löw (ed): *Christopher Polhems efterlämnade skrifter* vol. 4 (Uppsala, 1954 [1733]), 392.

encyclopaedists, craftsmen were more akin to the product of mechanics – that is, machines or *automata* – than to creative innovators in their own right.[116] Mechanical practitioners such as Duhre and Polhem had a similar relationship to craftsmen. On the one hand, they considered them to possess valuable knowledge that was the core of mechanical practice: knowledge that was indispensible for mechanically inclined boys. On the other hand, the craftsmen were radically different from the mechanicus and their teaching methods were unsuitable for mechanically apt boys.

This hierarchy between craftsmen and the mechanicus created an educational problem: how could a young future mechanicus learn the same skills as craftsmen without becoming identified by his contemporaries as a craftsman himself? Polhem considered solitary practice combined with the reading of books written by older mechanici and mathematici to be a possible alternative to an apprenticeship within the guilds: "for quick *ingenia* – for whom slavery does not become – beautiful books and descriptions are more suitable".[117] Duhre agreed with Polhem that experienced mechanical practitioners needed a way to communicate knowledge of mathematics and crafts to young boys in ways that guided them into useful manhood. His solution was a mathematical–œconomical educational institution, where young men would practise crafts integrated with mathematics. There, in a space ultimately controlled by Duhre, craftsmen would hand down their techniques to mechanically apt boys. In the workshops of this school, a site where mathematical education and practice of crafts were combined, the risk of becoming identified as a craftsman by learning crafts would be eliminated because it would be separated from the guild system and because the craftsmen would be subjugated to Duhre, an experienced mechanical practitioner.[118] For writers on mechanical education, to maintain a distance from the world of craftsmen was a means of establishing an order where their work was superior to mere manual labour.

---

[116] William H. Sewell: "Visions of labor. Illustrations of the mehcanical arts before, in, and after Diderot's Encyclopedie", in Steven L. Kaplan and Cynthia J. Koepp (eds): *Work in France. Representations, meaning, organization, and practice* (Ithaca, 1986), 275. On this double-speak of the encyclopaedists, see also Cynthia J. Koepp: "The alphabetical order. Work in Diderot's Encyclopédie", in Kaplan and Koepp (eds): *Work in France*, 243; Lorraine Daston: "Enlightenment calculations", *Critical inquiry* 21:1 (1994), 194. The reciprocal allegories between artisans and automata are also discussed in Schaffer: "Enlightened Automata".
[117] "för quicka ingenia som mindre ähro bequema till trähldom, tiänar bättre vackra böcker och beskrivningar"; Polhem: "Mechanica practica", 83.
[118] Duhre's school is discussed in more detail in Chapter 5.

Conclusions. The expected transgressions of useful manhood

In this chapter, we have found numerous narratives of the coming-of-age of a boy through exercises of mathematics and crafts. Given the mere number of reflections on the raising of mechanically apt boys, we can draw the conclusion that the man formed through these exercises hardly was an agent of change, acting unexpectedly in relation to his contemporaries. Instead, when older men imagined the coming of age of a mechanicus, they imagined a man defined by *expected useful transgressions*, continuously repeated and handed down from generation to generation. In these educational narratives, the exercises of mathematics and crafts complemented and improved one another. On the one hand, mathematics would nurture the boy's reason and imagination, so that he would not be stuck in the repetitive ways of craftsmen. On the other hand, the techniques of craftsmen would prevent a boy from getting lost and confused in a mathematical landscape of ideal curves and figures. The pure mathematician and the craftsman were imagined to be opposed extremes. They were, however, also seen as surprisingly similar. Exercised correctly, mathematics and crafts fostered a diligent man who contributed to the theocratic order; exercised incorrectly, they produced a confused man who would not meet the expectations of a manly subject. The useful life of a mechanicus was expected to be a *via media* between these extremes, which encompassed the expected virtues of a manly subject of the early modern Swedish state. In other words, it was imagined as a path from boyhood to manhood marked out in between a number of other personas.

The mechanicus was thus imagined to be both similar and different from craftsmen and pure mathematicians. He was expected to overcome each of their weaknesses, and to be useful to his state by mediating between their epistemologies and communities as well as by uniting reason and hard work. Sibum has shown how practitioners of *physica experimentalis* of the mid-eighteenth century inhabited a similar intermediary role. At this time, they were "seen as the ideal candidate to bridge theory and practice: the *third man*." To be recognised as inhabiting this position both enabled and disabled actors to perform in certain ways in relation to contemporary audiences.[119] As pointed out by Michel de Certeau, this third man "haunted enlightened discourse [...] and continues to do so today". However, the role of the modern engineer or "technocrat" is not identical to the role imagined by eighteenth-century writers. de Certeau sees these modern categories as an effect of a nineteenth-century process that "on the one hand isolated artistic techniques from art itself and on the other 'geometrized' and mathematized

---

[119] H. Otto Sibum: "Experimentalists in the Republic of Letters", *Science in Context* 16:1–2 (2003), 92.

these techniques."[120] In the narratives studied here, the relationship between mathematics and crafts is a more complicated matter. Although the process described by de Certeau might have been a result of the actions of mechanical practitioners, to see mechanics as a conscious act of mathematising crafts would be to confuse what is the historical cause here, and what is the effect. Although the mechanicus, as a third man, was presented as residing in between more established socio-epistemic categories, he was at the same time expected to transcend such categories. When mechanical practitioners presented the mechanicus as a third man, they thus expected him to be recognised as an *ideal* man.

These narratives of a *via media* should therefore not be seen a neutral descriptions of who early modern mechanical practitioners were. Instead, they should be analysed as performances in relation to the expectations of certain audiences. The fact that these texts presented the mechanicus as a mediator should not be taken to imply that mechanical practitioners could move effortlessly between the world of craftsmen and mathematicians. They might have been able to, in certain cases, but that is not what is at stake here.[121] What we *can* take from these narratives was that mechanics, as the combination of pure mathematics and crafts, was *imagined* by early moderns to reside in between a number of socio-epistemic categories. The coming-of-age narratives of the mechanically apt boy were stories of the making of men who contended perceived boundaries of scholars and craftsmen, in order to uphold the order in an early modern state. In other words: the mechanicus upheld and perfected the theocratic order by transgressing socio-epistemic categories recognised, by relevant audiences, as detrimental to it.

In early modern Sweden, permeated by patriarchal relationships, epistemic concepts such as *theoria* and *praxis* were part of an interconnected group of virtues, expected of a manly subject, which also included diligence and concord. When early modern authors imagined mechanics to be a path to manhood, it meant that they saw it as a means to acquire these virtues. In their narratives, older experienced mechanical practitioners imagined themselves to be motors driving a process of maturity that turned a boy into a mechanicus. Embracing this role in their publications, authors presented themselves as virtuous men and role models for potential students. They imagined themselves helping the public by recognising in a boy the seed of a man who was not yet there, and by making this boy in their own image. In their narratives, these authors claimed authority over boys' minds and bodies while they, at the same time, submitted and offered their services to the early modern state. These publications can thus themselves be seen as per-

---

[120] de Certeau: *The practice of everyday life*, 69.
[121] For a number of studies have pointed out the trangressive characteristics of mechanics and engineers. See page 20, note 23.

formances, or as the enactment of a range of tactics available to mathematical and mechanical practitioners, by which they, in relation to important audiences in the Swedish state, could make themselves into useful men.

In the coming chapters, we will repeatedly see how mechanical practitioners used similar tactics in order to shape themselves in relation to political power. Two results of this first chapter will be central in framing these following studies. First: the mechanicus should be understood in relation to the expected virtues of an early modern subject. Second: historical audiences expected mechanical performances to involve a well-defined set of transgressions. In the following, we will study this interdependence of transgressions and expectations in the relationships between mechanical practitioners and their audiences.

3. A geometric community takes shape

Men in a state are like musical instruments in an orchestra: they produce more or less agreeable sounds, depending on whether they are handled in a good or bad way.[1]

Mes pensées ou le qu'en dira-t-on, Laurent Angliviel de La Beaumelle (1752) quoted in *Rikets nytta af välbelönte ämbetsmän, tänkande informatorer och utvalde studerande*, Johan Niclas Zetherström (1765)

In early modern Europe, there existed many ways for a man to make himself an instrument of the early modern state. As shown in the previous chapter, experienced mechanical practitioners imagined that, by performing in relationships of superiority and subordination, a mechanically apt boy would become a useful young man. Their imagination highlighted the intergenerational aspect of learning mechanics, but it only provides one perspective on the mechanicus: that of experienced older men identifying boys with aptitude. From these narratives, it is hard to discern any institutional contexts of mathematics and mechanics. Also, the young man's perspective is not found there. What followed after a boy had become a man? What was he supposed to do and *where* was he supposed to be, once he had acquired at least some of the skills expected of a mechanicus? Also, what dreams did young men weave in relation to these imagined narratives of their elders?

This chapter focuses on young men who entered the Swedish civil service during the first half of the eighteenth century. More specifically, it studies the applicants to the Bureau of Mines [*Bergskollegium*].[2] The officials of the Bureau were responsible for overseeing two of the most important activities of the realm (next to agriculture): the production of, and trade with, metals. During the eighteenth century, metal production saw a marked increase, and it has been argued that this development can be attributed to the new rational methods put in place by cameralists, chemists and mechanical practitioners. I am less concerned with these changes. Instead, my main

---

[1] "Les hommes sont dans un Etat ce que des Instrumens de Musique sont dans un Orchestre: ils rendent des sons plus ou moins agréables, suivant qu'ils sont bien ou mal touchés."; Johan Niclas Zetherström: *Rikets nytta af välbelönte ämbetsmän, tänkande informatorer och utvlade studerande* (Stockholm, 1765).

[2] In the following I use Hjalmar Fors' translation of Bergskollegium into the "Bureau of Mines" and I translate *kollegium* into "bureau". For a longer discussion of this translation, see page 32, note 60.

concern here is the norms and skills of the community of men in the Bureau. Here, I argue that, during the first half of the eighteenth century, the Bureau developed into a community of trusted servants of the state, and that mathematics and mechanics gained a new and important role in this community.

Young men, in their late teens or early twenties, applied to the Bureau once they had reached adolescence. This chapter thus starts off where the previous one finished: with young men who, by merit of their education, could credibly present themselves as useful members of the state. Whereas the preceding chapter examined imagined intergenerational relationships formed around the performance of mechanics, here I study how performances of mathematics and mechanics were the basis of relationships in a community of men. I approach the Bureau as a community of practice, as brought forth by Jean Lave and Etienne Wenger. That is, I see it as a space for "situated learning" and to participate in its practices was to partake in a culture based on an "epistemological principle of learning".[3] My focus is on the reproduction of this community, through "legitimate peripheral participation", and on the social process of learning, by which a young newcomer was enculturated into the community and by which the community was reproduced.[4] Learning certain skills and norms was a way to align oneself to the Bureau's culture of knowing about and living in the world, in other words: its social epistemology. By the mid 1700s, mathematics became a key part of these processes, which formed both the Bureau and its officials.

Although the Bureau is the focus here, it is not my primary object of study. Instead, my interest is the performances of applicants and officials, and how such performances formed a community of men in the early modern Swedish state. The Bureau was *one* institution out of many, suitable for studying these relational performances. It was not the only state institution that was a community of mathematical and mechanical practice: mathematics and mechanics were also a part of the navy, artillery, fortification corps, Bureau of Commerce and Bureau of War. However, there are good reasons for focusing on the Bureau of Mines. Its archive contains an unprecedented wealth of sources that give insight into the expectations of the young men who shaped themselves together with the community they became a part of. Also, compared with the military, the applicants to the Bureau were a relatively homogeneous group. This makes it easier to identify broader groups and trends among them than among other large groups of young men

---

[3] Jean Lave and Etienne Wenger: *Situated learning. Legitimate peripheral participation* (Cambridge, 2005), 98.

[4] Ibid., 29. On learning as enculturation, see also Collins: "Learning through enculturation"; Collins: *Changing order*, 159. Collins' model of enculturation by and large corresponds to the model of Lave and Wenger. Most importantly, both see learning as a form of skill acquisition rather than as a communication of formal rules. Thus, in the text I use enculturation and "peripheral participation" interchangeably.

studying mathematics together (e.g., the volunteers in the fortification corps). Furthermore, in the Bureau one finds all the main actors of the following chapters: Christopher Polhammar, Emanuel Swedenborg and Anders Gabriel Duhre. Therefore, by understanding the learning situated there, one will also understand the making of these men, around whom the following chapters will revolve. The Bureau is therefore a good case for studying relational performances of mathematics and mechanics, and how young men were moulded in relation to the state to the expectations that constituted the mechanicus.

During the first half of the eighteenth century, numerous young men applied to auscultate (from the Latin verb *auscultare*, "to listen") in the Bureau. In other words, they wished to learn how to be mining officials, by participating in its work. Their letters of application are the main category of sources for the first section of this chapter. This material comprises 262 applications made by 237 applicants, scattered over the 118 folios of the Bureau's incoming communication from the years 1700–50. These applications constitute a previously unused source for understanding a broad group of men from the Swedish state. I read the applications as performances, by which the young men attempted to anticipate and adhere to the expectations of legitimate peripheral participants in the Bureau. I also read some of these letters against the protocols of the board (from 1700 to 1750, consisting of 68 folios). When this is the case, I use the protocols to piece together a more two-sided picture of becoming a part of the Bureau. I also combine the applications for auscultation with other requests – for example, for travelling, or for stipends.[5] Using these sources, I aim to delineate how the Bureau identified potential useful members, and correspondingly how young men presented themselves as such. What merits and characteristics did applicants include in their applications, and did they change between 1700 and 1750? How did the letters conform to a certain category of useful and virtuous civil servant? This section leads to the argument that, between 1700 and 1750, mathematics – and especially geometry – became an important part of the identificatory practices of a large group of applicants. By 1750, it was not only considered relevant for future mining mechanici, but held important symbolic function in forming a coherent community of civil servants.

In the second section, I turn to the accepted auscultators and the education they received within the Bureau. This section studies the role of mathematics as an exercise of enculturation, using handwritten manuscripts in mathematics, mechanics and natural philosophy copied by the auscultators together with commemorative speeches made by the Bureau's officials.

---

[5] The applications and protocols can be found in "Incoming correspondence to the Bureau of Mines", 1700–50, E4/105–223, Bergskollegiums arkiv, RA; "Protocols of the Bureau of Mines", A1/38–106, Bergskollegiums arkiv, RA.

Through these sources, I continue to explore the community the auscultators were enculturated into, how this enculturation came about, and the meanings mathematics – especially geometry and mechanics – had in the establishment of the Bureau's community of civil servants.

The mining sciences

Mining was of great importance for the early modern Swedish economy. Consequently, during this period, the Swedish state took an increasing interest in metal production. In the late sixteenth and early seventeenth centuries, Swedish mining was administered by the fiscal chamber, and the mining administration was not a formally regulated governmental body. From the 1630s, however, the government took measures to improve the supervision of the mines. In 1637, Queen Christina founded a general mining board [*generalbergsamt*] that inspected mining and smelting. The statutes stated that its officials should diligently police the mines of the crown and that the board should recruit competent men who could replace dismissed or deceased officials. Furthermore, these statutes imposed some expectations on the mining officials: they were not allowed to own parts in mines, they should be satisfied with the salary given to them as civil servants, and they were not allowed to travel between mines "for private gain".[6] The general mining board soon consolidated its place in the state. As early as 1649, it attained the status of an administrative bureau. From this point on, the council was not merely a board of experienced men, but a Bureau of Mines.[7]

By the early eighteenth century, the Bureau had become a well-established institution, where the policing of metal making was interwoven in bureaucratic practice aiming at revenue for the state. There, administrative sciences, such as jurisprudence and œconomy, coexisted with other knowledge geared towards metal production, which the officials called mining sciences [*bergssciencer* or *bergswettenskaper*].[8] These sciences were not only a way to categorise the activities of the Bureau's officials as a form of

---

[6] "för privat nytta"; The statutes are quoted in Jan-Olof Hedström: *-igenom gode ordningar och flitigt upseende-. Bergsstaten 375 år* (Uppsala, 2012), 13–16, quote from 15.
[7] On the importance of the Swedish metal production, see Göran Rydén: "Skill and technical change in the Swedish iron industry, 1750–1860", *Technology and culture* 39:3 (1998), 383; For an analysis and overview of the development of Swedish metal making regions, see Anders Florén and Göran Rydén: *Arbete, hushåll och region. Tankar om industrialiseringsprocesser och den svenska järnhanteringen* (Uppsala, 1992), 98–102, 124–8. For an overview of the early Bureau of Mines, see Lindqvist: *Technology on trial*, 95–6; Mirkka Lappalainen: "Släkt och stånd i Bergskollegium före reduktionstiden", *Historisk tidskrift för Finland* 87:2 (2002), 140–72; Fors: "Kemi, paracelsism och mekanisk filosofi", 172–3; Fors: *The limits of matter*, 46–7.
[8] Fors: *The limits of matter*, 7. Importantly, as pointed out by Fors, "the pursuit of science of philosophy was never a primary concern of the Bureau as a whole"; ibid., 7 note 22. The mathematical community discussed in this chapter should thus be seen as synonymous with the while Bureau.

knowledge making, but were also a means of status. In controversies concerning their rank, the men of the Bureau repeatedly used their double competence – of juridical matters as well as of the so-called mining sciences – as an argument for why they could claim a higher status, compared with other servants of the state.[9]

What constituted the mining sciences was historically contingent. Over time, they came to consist of, for example, mineralogy, chemistry and experimental physics.[10] As shown by Fors, in the late seventeenth century, Urban Hiärne introduced chemistry into the Bureau's work in a *Laboratorium Chymicum*.[11] Similarly, from the 1690s, practical mechanics found a place in the Bureau. In 1697, Christopher Polhammar, with help from his main patron Fabian Wrede (the president of the State Office [*Statskontoret*] and the Bureau of Accounts [*Kammarkollegium*]), managed to convince the king, Karl XI, to establish a *Laboratorium Mechanicum* under his command. This institution was formed on the model of Hiärne's chemical counterpart.[12] Both facilities were established by royal command and were under direct personal control of their directors.[13] In Polhammar's *Laboratorium*, young men were to be educated in mechanics and mathematics. To encourage such apt young men, on 22 July 1699, two stipends in mechanics were established in the *Laboratorium*.[14] Fors has argued that the establishment of a mechanical laboratory was part of a shift in the Bureau from curious chemical knowledge to useful knowledge based on mechanics, and that the subsequent king, Karl XII, showed more interest in mechanical projects related to the Swedish war efforts than in transmutive chemistry. Polhammar's laboratory can be seen as a manifestation of these changing priorities of the early eighteenth century, when mechanics replaced chemistry as the science favoured by the state.[15] However, by 1715, the activity in the *Laboratorium Mechanicum* dwindled. Lindqvist has argued that between 1715 and 1725 "mechanical engineering was [...] in poor shape" in the Bureau, while Polhammar was occupied with other issues brought about by a new personal relationship with Karl XII.[16] Interestingly, it was now that explanations of nature based on geometry and mechanical philosophy took hold in the Bu-

---

[9] The Bureau of Mines to The Royal Majesty: "Concerning rank", 1719-06-01, 8/19, Brev från kollegier m.fl. [...] till Kongl maj; Bureau of Mines to The Royal Majesty: "Concerning rank", 1730-01-26, 8/22, Brev från kollegier m.fl. [...] till Kongl maj.

[10] Widmalm: *Mellan kartan och verkligheten*, 60.

[11] Fors: *The limits of matter*, 48–52.

[12] Ibid., 82.

[13] Lindqvist: *Technology on trial*, 99.

[14] Ibid., 97–8; On the stipends, see: Carl XII: "Decision on instating a stipendio mechanico", 1699-07-22, E 1/7/594, Bergskollegiums arkiv, RA.

[15] Fors: *The limits of matter*, 71. Urban Hiärne as a client of Karl XI, and as a "royal chemist", see Fors: "Kemi, paracelsism och mekanisk filosofi", 173–4. Still, as we will see in Chapter 4, the king's perceived interest in mechanics was as much *a result* of his patronage of Polhem, as what motivated their relationship.

[16] Lindqvist: *Technology on trial*, 99. For a detailed analysis of this relationship, see Chapter 4.

reau and when chemistry was submitted to a geometric and mechanical framework. Thus, from the 1720s, a new form of mechanical and geometric chemistry started to gain a place in the Bureau.[17]

A young man entering the Bureau in 1750 entered a community where mathematics carried connotations different from those it had held at the turn of the century. When the place of mathematics in the mining sciences shifted, the officials did not only start to perceive nature in a new way – the men in the Bureau also came to relate differently to each other. Instead of being intimately linked to mining mechanics, machine building and Christopher Polhammar, by the 1720s, mathematics became established as a foundation of the Bureau's community of officials, their work and their knowledge of mining. At this time, mathematics also became central in the forming of a large group of young aspiring officials into civil servants who based their exercise of public power on a mathematical foundation.

Auscultation. The Bureau of Mines as a place of education

The Bureau, like all other communities, did not reproduce effortlessly. As pointed out in the previous chapter, in spite of the attempts by the civil administration to reform academia, Swedish universities kept their strong ties to the church well into the eighteenth century. In order to satiate the demand for specialised state officials, the bureaus instead developed an auscultatory system.[18] The auscultatory system of the Bureau of Mines – like those of the other bureaus in Stockholm and of the juridical system – developed in the second half of the seventeenth century. Auscultation – in its basic form the right to listen in to the meetings of the board, but generally also involving participation in the Bureau's work – could be seen as a form of legitimate peripheral participation. It was a process of learning-by-doing, by which the auscultators developed into officials and the Bureau reproduced itself. During the seventeenth century, the small scale of the Bureau's educational efforts, and the personal relationships between students and teachers, resembled not only the imagined intergenerational relationships discussed in the previous chapter, but also the master–apprentice relationships found in guilds. By the beginning of the eighteenth century, this would change, because the Bureau's system of auscultation would grow dramatically.[19]

---

[17] Fors: *The limits of matter*, 90–2.

[18] Gaunt: *Utbildning till statens tjänst*, 31.

[19] Auscultation was a system of learning through participation in various forms of work, and thus it resembles the apprentice systems of the guilds. These similarities have been discussed by Fors: *The limits of matter*, 95. For a discussion of early modern mining education in a European context, see Donata Brianta: "Education and training in the mining industry, 1750–1860. European models and the Italian case", *Annals of science* 57:3 (2000), 267–300; For a discussion of how mining

From the early 1650s, the Bureau had two auscultators, who were considered high officials: in the list of officials in the Bureau they were named after the members of the board, and before all other minor civil servants. By a royal decree on 4 May 1673, these first two auscultators reformed into extraordinary and salaried assessors with a right to speak on the board.[20] In 1686, Karl XI reinstated two auscultator positions. These auscultators would, however, come to differ from those of the mid seventeenth century. In a letter of 1691, from the Bureau to the king, the board argued that the auscultators should be considered holders of stipends. The board stated that the auscultators were supposed to be young men who wished to learn mining, and if they were given permanent positions it would hinder the necessary flow of new men into the Bureau.[21] As pointed out earlier, in 1699 the Bureau established two additional stipends, linked to Christopher Polhammar and his *Laboratorium Mechanicum*, that were explicitly geared towards mechanics.[22]

When the Bureau made the salaried positions of auscultator into stipends, they also uncoupled auscultation from financial compensation. The stipends became considered a *beneficium* given to the most worthy of young men, but the Bureau also came to accept an increasing number of auscultators who did not receive any form of monetary compensation. By the turn of the century, auscultation was thus a two-tier system consisting of a small number of promising men with stipends (*beneficium ascultandi*) and another group of unpaid auscultators (*venia auscultandi*). In a short history of the auscultatory system written at mid-century, the official Daniel Tilas told of how the group of *venia auscultandi* had grown as time had gone by. In Tilas' view, this group of unpaid young men might even have been too large at times.[23]

Whereas the number of auscultators remained relatively small throughout the seventeenth century, by the 1690s the system started to expand. Although the number of holders of stipends doubled in 1699, it was the establishment of a second tier of auscultators that accounted for the main expansion of the Bureau's educational system. From 1670 to 1689, the Bureau only accepted 10 auscultators in total into its ranks; in the decade of the 1690s alone, it accepted 27. The period of the Great Nordic War in the early eighteenth century saw a small decline in the enrolment of auscultators. Still, during the first two decades of the eighteenth century (1700–19),

---

education in Sweden differed from that of other European states, and especially Britain, see ibid., 280–1. On the similarities between Leve and Wenger's concept of legitimate peripheral participation and apprenticeship, see; Lave and Wenger: *Situated learning*, 29–34.

[20] Johan Axel Almquist: *Bergskollegium och Bergslagsstaterna 1637–1857. Administrativa och biografiska anteckningar* (Stockholm, 1909), 16–17.

[21] The Bureau of Mines to Carl XI: "On stipends to the auscultators", 1691-06-16, vol. 7, Kollegiers m.fl, [...] skrivelser till Kungl Maj:t, RA.

[22] Carl XII: "Decision on instating a stipendio mechanico".

[23] Daniel Tilas: "Om nuvarande Bergskollegium, dess historia, organisation, verksamhet m.m."; D6/14/225–305, Bergskollegiums arkiv, RA, 270.

37 auscultators were accepted into the Bureau. By the end of the war, the number of auscultators had grown rapidly, with 35 accepted in the 1720s, 82 in the 1730s and 88 in the 1740s. Consequently, the educational structure of auscultation hardly remained unchanged over this time period.[24]

Lindqvist has studied the officials of the Bureau over the brief period of 1715–25 and has sketched the "average [Bureau] of Mines official in the eighteenth century" using biographical details of assessors from 1720–1815. The average career of an assessor began with studies "at Uppsala University until he was 20 years old"; then he applied to be an auscultator and after being accepted he travelled abroad.[25] Similarly, Fors has argued that auscultators could look forward to "rather good career prospects," since "the fifty assessors who were appointed to the Bureau's governing board during the years 1708–94, all but three had begun their career at the Bureau as auscultators."[26] But of course, just because most assessors had been auscultators, it does not follow that most auscultators became assessors. Also, it is not evident that the career of an assessor was the career of an average official. To become an assessor was the high point of a very successful career, and assessors are thus not necessarily representative of a broader set of officials or auscultators. By focusing on the young men approaching the Bureau, we get a better understanding of the expectations of the broad group of auscultators than if we sketch the career of assessors in retrospect. Through these applications we can see how contemporaries saw the boundaries to participation within the bureaucratic community, and how these boundaries changed over time.

Unsurprisingly, neither of the two tiers of auscultation was open to everyone. In the early 1720s, only about half the applicants were accepted into the Bureau.[27] In order to be seen as a legitimate peripheral participant, the would-be auscultator had to present himself in a written letter. There he stated his reasons for applying, listed his merits and often also included letters of recommendation. As the auscultatory system grew, the Bureau saw

---

[24] Almquist: *Bergskollegium och Bergslagsstaterna 1637–1857*, 116–17. These numbers show that Lindqvist's brief study from 1715–25, indicating that on average "three or four Auscultators were admitted each year", is not representative for the rest of the eighteenth century or for the 1600s; Lindqvist: *Technology on trial*, 101. Furthermore, it shows that the scale of the Bureau's educational system changed drastically from the mid-seventeenth to the mid-eighteenth century.

[25] Lindqvist: *Technology on trial*, 101. Of course, being an average, the picture Lindqvist sketches ignores many of the ways the ideal official changed over the eighteenth century. But furthermore, by focusing on the assessors what Lindqvist displays is hardly an average man, but one route out of many. Of course, the majority of auscultators did not continue to become assessors.

[26] Fors: *The limits of matter*, 84.

[27] Lindqvist: *Technology on trial*, 101. Because the Bureau of Mines never compiled a list of applicants that were not accepted, it is difficult to assess the proportion of accepted auscultators in a longer time period. But judging from the Bureau's protocols, it seems as if the relative number of dismissed applicants was exceptionally high during the time period that Lindqvist studies. This is unsurprising, as the economy of the Swedish state was devastated during the last years of the Great Northern War and the first years after the peace.

an influx of such applications. The applications were structured relatively homogeneously, which could be interpreted as the applicants being aware of, and anticipating, certain requirements from the Bureau. All applications can be seen as having been written with the explicit purpose of being accepted into the Bureau. Consequently, the applications are an interesting genre of self-fashioning. Using the applications, we can follow how the rules for legitimate peripheral participation changed over the first half of the eighteenth century, and how these rules shaped the applicants and their applications even before they were accepted.[28]

The applications pose a methodological challenge: individually each application tells us about the dreams, aspirations and merits of early eighteenth-century men hoping to enter public service. However, in order to study the dialectic process through which the officials were enculturated into, and simultaneously changed, the Bureau, we must go from these individual texts to an aggregate level. How then do we go from individual self-presentations to historically contingent personas found in between the individual and the collective? In order to understand the content of the letters at this intermediate level, I have coded a number of tropes found in the various texts (see Figures 1 and 2) and compiled them by decade. These tropes are not simple keywords, but rather specific topics that I have identified as prevalent in the letters. By studying the trends in this dataset, and by relating the whole set to individual texts, we can delineate changes and continuities in the self-presentations over the studied time period. This aggregate level is thus not an end in itself, but a means to grasp the demands the applicants anticipated from the community, and in relation to which they acted.

Applications to auscultate as self-presentation and identification

In the light of the discussion in Chapter 2 on the relationship between the state administration in Stockholm and the university in Uppsala, one might assume that the purpose of the system of auscultation was to sidestep universities, which resisted radical changes. While this explanation is partly correct, matters were somewhat more complicated and cannot simply be understood as a power struggle between two separate entities. In the applicants' recommendations from academia, one can see how the Bureau and Uppsala University were tied together over the first half of the eighteenth

---

[28] The numbers and the analysis in this section, unless otherwise stated, are based on my analysis of the applications found in the incoming letters of the archive of the Bureau of Mines from 1700–50 (RA, Bergskollegiums arkiv, E4/159–223).

century.[29] In the first decade of the 1700s, only 20 % of the applications mentioned academic merits. But already in 1710, it became more or less the norm for applicants to mention their university education (74 %). With a dip in the 1720s (43 %), just after the Great Northern War, the figure then stabilised to around 75 % in the 1730s and 1740s (see Figure 2 and 3).[30] Simultaneously, Uppsala University established itself as the major university for educating future auscultators. From the 1700s to 1750 – when an official mining exam was introduced in Uppsala – the percentage of applicants mentioning studies in Uppsala rose from 20 % to 67 %. Thus, as the number of auscultators grew rapidly, the number mentioning academic studies increased and the number explicitly mentioning Uppsala rose about as much (while the absolute numbers from the other two universities in Sweden only increased marginally). One conclusion from this increase in numbers would be that the establishment of a civil mining exam in Uppsala in 1750 was recognition of the fact that the university had already developed into a semi-official school for future mining officials by the 1730s and 1740s.[31]

In anticipation of the demands of an academic background, it was common for auscultators to acquire a basic academic training, preferably in Uppsala, before applying. For the applicants, an academic degree was not only proof of a social background necessary for serving the state, it also gave witness to basic experience and competence, seen as the basis for the Bureau's work. The only applicant over the whole period who explicitly discussed his *lack* of academic experience was Per Fahlander (1719–96), who applied in 1750. At a time when academic studies were about to become an official requirement, Fahlander made excuses for his lack of such credentials. He argued that his mathematical inventions were as good evidence of mathematical competence as a university degree:

---

[29] For example, Fors has pointed out how professors in Uppsala used contacts in the Bureau as a resource for paronage. The connections between Uppsala and the Bureau have been discussed in Fors: "Kemi, paracelsism och mekanisk filosofi", esp 189; Fors: *Mutual favours*, 46–9; Fors: *The limits of matter*, 84. This process, by which professors and bureau officials were tied together, can be understand using the concept of "patronage couples" by Hodacs and Nyberg: *Naturalhistoria på resande fot*, 125.

[30] In comparison, only thirteen mention Lund and three Åbo, at the time the two smaller universities in the Swedish realm. The number of auscultators who mention the smaller universities is more or less constant over the studied time period. Lindqvist is thus correct in his statement that the typical official of the Bureau of Mines between 1720–1815 had studied in Uppsala; Lindqvist: *Technology on trial*, 101. However, it should be added that this tendency was reinforced over the first half of the eighteenth century. This finding also corresponds to the view of Swedish civil sevants, as discussed in Cavallin: *I kungens och folkets tjänst*, 159–163.

[31] Almquist: *Bergskollegium och Bergslagsstaterna 1637–1857*, 46.

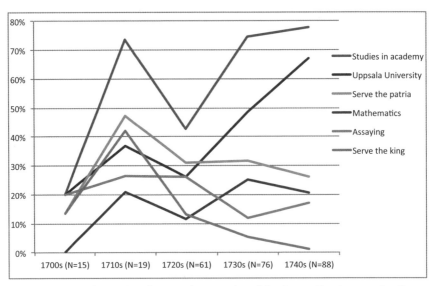

Figure 2. Proportions of various topics mentioned in the applications to the Bureau of Mines between 1720 and 1750

Occurrences of topics in applications to auscultate 1720–50						
	1700s	1710s	1720s	1730s	1740s	1720–50
Topics in applications						
Total number of applications	**15**	**19**	**61**	**76**	**88**	**244**
Mathematics	0% (0)	21% (4)	11% (7)	25% (19)	20% (18)	18% (44)
Physics	0% (0)	16% (3)	10% (6)	9% (7)	6% (5)	7% (18)
Assaying	20% (3)	26% (5)	26% (16)	12% (9)	17% (15)	16% (40)
Mechanics	7% (1)	21% (4)	3% (2)	11% (8)	3% (3)	5% (13)
The law	7% (1)	16% (3)	20% (12)	8% (6)	8% (7)	10% (25)
Mentions father/parents	13% (2)	21% (4)	15% (9)	9% (7)	8% (7)	9% (23)
Serve the patria	13% (2)	47% (9)	31% (19)	32% (24)	26% (23)	27% (66)
Serve the king	13% (2)	42% (8)	13% (8)	5% (4)	1% (1)	5% (13)
Uppsala University	20% (3)	37% (7)	26% (16)	49% (37)	67% (59)	46% (112)
Åbo University	0% (0)	0% (0)	2% (1)	3% (2)	0% (0)	1% (3)
Lund University	0% (0)	0% (0)	5% (3)	7% (5)	6% (5)	5% (13)
Interest [*lust*]	40% (6)	47% (9)	23% (14)	28% (22)	30% (28)	26% (64)
Studies in academy	20% (3)	74% (14)	43% (26)	75% (57)	78% (70)	63% (153)
Sciences related to mining	13% (2)	26% (5)	21% (13)	28% (21)	34% (30)	26% (64)
Combinations of topics						
Mathematics + physics	0% (0)	11% (2)	8% (5)	8% (6)	6% (5)	
Mathematics + mechanics	0% (0)	21% (4)	2% (1)	8% (6)	2% (1)	
Assaying + mathematics	0% (0)	0% (0)	2% (1)	3% (2)	5% (3)	
Assaying + mechanics	0% (0)	0% (0)	0% (0)	1% (1)	0% (0)	
Assaying + physics	0% (0)	0% (0)	2% (1)	0% (0)	1% (1)	

Figure 3. Table of topics in auscultation applications 1700–50, grouped by decade; relative (and absolute) values

I humbly presume that the displays of competence in *mathematical* knowledge, which I have already provided during my period of service, should be considered sufficient evidence of the *experience* I have therein, as if I had undergone an academic examination.[32]

The applicants generally described their academic education as a means to an end. Often this end was to gain useful skills, which could be used within the mining sciences and production. Peter Westerberg, who applied in 1725, described how he had "established a foundation in *studies* and useful knowledge at the *academy* in *Uppsala*, and now is ready to seek further instruction on what could eventually make him serviceable in carrying out some public position".[33] Likewise, the applicant Carl Leijell in 1735 expressed how he "because of a particular interest in mining, as a part of his academic studies had endeavoured to acquire knowledge of *physics* and *mathematics*".[34] By presenting their academic merits, the applicants expressed their suitability for a public role as well as their aptitude for certain recognised forms of knowledge promoted by the Bureau.

In the applications, auscultation was commonly referred to as a way of learning, and thus consequently also as a process of enculturation. The applicants described participation in the Bureau as a means to learn more about the mining sciences, while simultaneously becoming a patriotic part of the state. In 1725, Jonas Schnack wrote that he wished to auscultate in order to "make [himself] so much more competent and skilled in [mining]". Gustav Littmarck (1719–84), in 1737, wished to auscultate in order to "have the opportunity to continue to acquire what is needed, to in time make myself skilled and worthy of […] advancement."[35] These descriptions of the Bureau as a place of learning went hand in hand with proclamations of the Bureau as a benefactor of useful knowledge. Among many others, Johan Camitz in

---

[32] "de prof som iag redan under min tienste tijd afgifwit på insickt utij *mathematiske* wettenskaper förmodar iag i all ödmiukhet läre kunna anses för tillräckeliga bewijs, om then *erfarenhet* iag der utinnan ger, som om iag hade undergåt *academisk examen*"; Per Fahlander: "Application to auscultate", 1750-11-12, E4/223/171, Bergskollegiums arkiv, RA. The applicants to the Bureau generally did not mark the date of writing their letters. However, the officials in the Bureau noted the date of receiving them. Here, and in the following, I refer to the letters to the Bureau using this date.
[33] "wid *Academien* i *Upsala* lagt någon grund uti *Studier* och nödiga wettenskaper, och nu mera är sinnad at söka widare underrättelse om det som i framtiden kan göra mig tienlig till någon *publique* tienst förträdande"; Peter Westerberg: "Application to auscultate", 1725-02-16, E4/156/187, Bergskollegiums arkiv, RA.
[34] "under en besynnerlig och [sic] lust till Bergswäsendet ibland andra *Academiska Studier* sökt inhämta the wettenskaper, som höra til *Physiquen* och *Mathematiquen*"; Carl Leijell: "Application to auscultate", 1735-12-20, E4/178/110, Bergskollegiums arkiv, RA.
[35] "så mycket mera uti dessa saker må göra mig *habil* och skickelig", Jonas Schnack: "Application to auscultate", 1725-10-26, E4/156/189, Bergskollegiums arkiv, RA. "hafwa tillfälle at kunna widare inhämta hwad som fordras, at med tiden giöra mig skickelig och wärdig till Eders *Excellences* samt Kongl. May:ts och Riksens Höglofl. Bärgz *Collegii* nåd och höggunstige befordran."; Gustav Littmarck: "Application to auscultate", 1737-03-21, E4/184/223, Bergskollegiums arkiv, RA.

1741 pointed out that the Bureau held a "highly praised affection for the education of young people in honourable and useful knowledge".[36] Similar descriptions of the Bureau as a benefactor of useful knowledge related to mining can be found in many of the applications from that time.

Still, the knowledge that applicants referred to by writing of "useful knowledge" shifted over the studied period. As seen in Figure 3, already at the turn of the century, applicants commonly mentioned their competence in the art of assaying (20 %). By the 1720s, many applicants also mentioned a legal education (20 %).[37] Likewise, from the 1710s, it was common for applicants to mention having studied mathematics. Examples of useful fields of knowledge for an applicant can be found in "A short report on how a young man, who wish to seek work in the mining industry, should spend his time." This handwritten document, written or copied by an auscultator of the Bureau, described one perspective on the ideal education of a mining official. Up to the age of 15 he should study "*Studia humaniora*" (i.e., a classical education consisting of e.g. grammar, rhetoric and poetry). Then, after matriculating at a university, he should devote himself to studies related to mining: "geometry, physics, mechanics and hydrostatics." At the same time, he should study "the law and the *process*" and he should "practise drawing." At the age of 19 or 20, he should apply to auscultate in the Bureau.[38] To present one's grasp of such knowledge was to present one's aptitude for the mining sciences.

Emotion, kinship and childhood

Another common way of arguing for acceptance into the Bureau's community was to describe an emotional connection to mining and the mining sciences. Many applicants said that an inclination for mining had guided their studies and exercises from a young age. These applicants discussed mining in ways similar to the narratives of how a mechanicus was defined through the boy that he had once been, discussed in the previous chapter. In the same way, descriptions of an interest in mining, dating back to childhood, reflected well on the young man who applied. Anders Barchaus, who applied in 1729, stated that it had "always been my greatest desire to learn the knowledge of mining". Similarly, in 1735, Lars Fredric Bonde wrote of

---

[36] "Kongl. Collegii högtbepriseliga ömhet för ungdommens förkåfring uti anständiga och nyttige wettenskaper"; Johan Camitz: "Application to auscultate", 1741-11-03, E4/196/273, Bergskollegiums arkiv, RA.

[37] The applicants mentioning the law and assaying can be divided into two more or less distinct groups. Only one applicant in the 1720s, Christian Georg Danckwardt (1701–63), mentioned both assaying and the law in his application.

[38] "*Studia humaniora*"; "*Geometrien, Physiquen, Mechaniquen* och *Hydrostatiquen*."; "lagen och *Processen*"; "Öfwa sig i Ritande."; Lars Schultze: "Kort betänkande huru en yngling som tänker söka sin fortkomst vid bergsväsendet bör sin tid anlägga", D 1433, UUB. For a discussion of the collection where this text is found, see below.

how "ever since my earliest youth, I have carried an especial interest for mines". Daniel Tilas instead described how he discovered his interest for mining during his time in Uppsala.[39] Such expressions of enthusiasm and interest made the applications as well as the discussions of the mining sciences into more than a mere list of competences or contacts: many applicants evidently also anticipated the need to display their emotional connection to the Bureau. By presenting such a connection, applicants could show that they possessed, or aspired to, a certain diligent character, which defined a manly subject and a useful civil servant.

Declarations of kinship and place of birth are two other reoccurring tropes in the applications. About one-tenth of the applicants explicitly referred to their father or to their parents. In a few cases, especially when the auscultator was very young, it was even the father who wrote the application for their son.[40] Many applicants named old or even dead relatives. In 1723, Carl Urban Hiärne, the son of the previously mentioned Urban Hiärne, argued that, if he were accepted as an auscultator, it would make his father happy, who all his life had provided his service to the Bureau. Other applicants, with a less prominent heritage, also referred to kinship. For example, in 1735, Per Strömmer Gyllenhök hoped that his late grandfather's merits would also reflect well upon him. Peter Wallström pointed out, in 1743, how he, between university terms, had lived with his father, who was "the inspector of the ironworks of Axmar".[41] By discussing kinship, these men showed how they were part of family networks that traversed the Bureau as well as the Swedish metal production.

Although the Bureau was dominated by a small number of families, declarations of kinship were less common than one might assume.[42] Possibly, surnames were a more discrete signifier of kinship than explicit references. Nevertheless, kinship and place of birth interplayed in many applications: to present your family was also to present your origin. Many applicants, who were born in Falun or in other mining districts, highlighted their

---

[39] "altid min Högel:e åstundan warit att utij de till Bergwäsendet hörande wettenskaper mig *evertuera*, som jag och mig derom på det möjeligaste biflitat"; Anders Barchaus: "Application to auscultate", 1729-11-23, E4/163/758, Bergskollegiums arkiv, RA. "Alt ifrån min späda ungdom har iag haft till Bergswärk en särdeles lust"; Fredric Lars Bonde: "Application to auscultate", 1735-10-13, E4/178/114, Bergskollegiums arkiv, RA; Daniel Tilas: "Application to auscultate", 1732-03-22, E4/178/1, Bergskollegiums arkiv, RA.
[40] The proportion of applicants discussing their parents was relatively constant 1700–50, but decreased slightly over time (see Figure 3). From 1720–50 five men applied on behalf of their sons: Nils Reuterhold (1737), Eric Sohlberg (1737), Nils Söderhjelm (1749), Christopher Risell (1742) and Samuel Troili (1745).
[41] "Inspectoren på Axmars Bruk"; Carl Urban Hiärne: "Application to auscultate", 1723-01-23, E4/150/260, Bergskollegiums arkiv, RA; Gyllenhök Per Strömmer: "Application to auscultate", 1735-02-28, E4/178/10, Bergskollegiums arkiv, RA; Peter Wallström: "Application to auscultate", 1743-05-09, E4/205/32, Bergskollegiums arkiv, RA.
[42] Lindqvist: *Technology on trial*, 218. Fors points out how family networks were central to the close links between the Bureau and Uppsala University, Fors: *The limits of matter*, 84.

geographical origin directly. They pointed out that they had been brought up in proximity to mining activities, where they had also acquired their early interest in mining. Discussions of kinship and places of birth were yet another way of using one's childhood to present oneself as a useful man.[43] Duhre expressed a similar view in a publication of his. He pointed out that the students with the most aptitude for the mining sciences were those "born in the mining districts, who from their childhood have had opportunity to see and to familiarise themselves with such work, which become and befit a man experienced in mining and smelting."[44]

These applications, written by young men from the mining districts, described a purposeful path. Being born in areas of relevance to the Bureau, they had gone to the universities with the explicit purpose of continuing on to the Bureau and to mining.[45] Kinship, place of birth and inclination thus converged in exclamations of a professed interest in and aptitude for the mining sciences. Through such tropes, the applicant presented himself as having experience of and competence in mining, despite his young age. Declarations of kinship and origin were interpreted as proof of well-established habits of diligence and hard work, brought about through good upbringing and socialisation in the circles of mining officials.[46] Thus, the applications could be seen as mirroring the narratives studied in the previous chapter. In other words, the applicants anticipated the expectations of the Bureau by presenting their lives in accordance with the imagined coming-of-age narratives of their elders.

---

[43] Similarly Mirkka Lappalainen has pointed out how, in the seventeenth century, the fact that the Bureau was permeated by certain noble families was not interpreted as contradicting the wish that it should be populated by competent men; Lappalainen: "Släkt och stånd i Bergskollegium före reduktionstiden", 170–3. Compare also to Svante Norrhem: *Uppkomlingarna. Kanslitjänstemännen i 1600-talets Sverige och Europa* (Umeå, 1993), 66–8.

[44] "anses för de skickligaste, at lära det som angår Bergwärcken, hwilka födde uti Bergslagerna ifrån Barndomen hafwa haft tilfälle at se och umgå med sådana sysslor, som en förfaren Bergsman pryda och anstå."; Duhre: *Förklaring*, 25–6. A similar declaration of the place of birth for a man in the Bureau can be found in Anders Berch: *Åminnelse-tal, öfver commerce-rådet Henric Kalmeter, efter kongl. vetensk. academiens befallning, hållit i stora riddarhus-salen den 13. martii, 1752* (Stockholm, 1752), 6–7.

"Bergsman", which denotes specific men of early modern metal production, has not equivalent in English; Fors: *Mutual favours*, 22–3; Fors: The limits of matter, 14–5. The term can roughly be translated into "a man working with mining and smelting".

[45] 29 out of 192 applications describe such an interest from a young age. 13 applicants describe their upbringing near mines as a way of becoming interested in and knowledgeable of mining sciences.

[46] Lappalainen has pointed out the lack of research on kinship networks and social frameworks of lower level officials of the Bureau. Lappalainen: "Släkt och stånd i Bergskollegium före reduktionstiden", 169. For a more detailed study on these matters, the applications to auscultate could be an important starting point.

Application as submission

The acts of pious submission, common in the narratives of the previous chapter, can also be found in the applications to the Bureau. There, applicants presented themselves as inclined to study useful sciences out of patriotic sentiment and the Bureau was described as an institution where knowledge and learning were interwoven with diligent work for the good of the *patria*. In the applications, desire to learn the mining sciences was repeatedly tied to the process of becoming a servant of the *publicum* (i.e., of the state or the sovereign). In eighteenth-century Sweden, to be a civil servant was to be a good citizen – a *patriot* – and this was the requirement that the applicants anticipated in their applications. As already pointed out in Chapter 1, the concepts of virtue and usefulness were interdependent at this time. To be a useful and virtuous subject was to be a pious man, and being a civil servant was one way to be recognised as such.[47] Because they were submitted to the state, the acts of men who embraced the civil culture of the bureaus were generally interpreted as free of private interest. As long as these men met the obligations of a man of the state – a *publicus* – possibilities of state support for projects, as well as opportunities for stipends and paid positions, opened up for them. On the other hand, as we will see in the coming chapters, when the performance of selflessness failed, the consequences could be severe.

According the linen manufacturer Samuel Crispin Ulff, in a proposal for governmental support for his manufactory sent to parliament in 1729, the opposite of a public servant was the man who served no one:

> Such people, who only serve themselves and no one else, are the burden, weight and trouble of the public. They have severed themselves from the burghly state of the *republic*, not unlike those, who worship themselves, and who believe in their own skill, power and ability. The just punishment of these men is that in coming distress and misfortune they should be left alone, and should never enjoy any of the good found under the wings and protection of the *publicum*.[48]

For Ulff, the *publicum* was a superior but caring patriarch and he accredited trust to actors who were integrated into a system of superiority and subordination and who were therefore sheltered by the wings of state

---

[47] On virtue, manhood and piety, see page 42–47. On virtue and manliness, see also Rimm: *Vältalighet och mannafostran*, 42. For an analysis of the concept of passions, usefulness and virtue in Swedish eighteenth-century economic literature, see Runefelt: *Dygden som välståndets grund*, 9.

[48] "Slikt Folck, som tiena sig allena och ingen annan, äro det Allmännas Last, Tynga och Beswär, de der sielfwa afstympat sig ifrån det Borgerliga Wäsendet i *Republiquen*, ey olykt dem, som dyrcka sig sielf, och tro på sin egen Skickelighet Mackt och Förmåga. Sådane, är deras egit Straff rättmätigt i påkommande Nöder och Olyckor, at de böra lemnas sig sielfwa och aldrig niuta något godt under *Publici* Skygd och Wingar"; Samuel Crispin Ulff: *En liten doch utförlig grundritning och handledning, til de metall- och linne-manufacturier, som nu inrättas i Hälsinge-land, at drifwas på åtskilligt nytt arbetz sätt, särdeles af landetz egne producter; hwilka under twenne societeter komma at sortera, såsom deras ägare och förläggare: blifwandes här ock något förestält om manufacturerne i gemen* (Stockholm, 1729), 23.

power.[49] In the applications, we find these tropes of the ideal civil servant as a selfless patriot expressed by a large number of young men. It was common for applicants to use tropes of patriotism or the public good from the 1710s onward. Hence, the applications adhere to the "utilist" ideals of science that Sven Widmalm has found in the late Age of Liberty, but already existed in the late Caroline era.[50] Also, from the 1710s to the 1740s, the applications that expressed a wish to serve the king dropped from 42 % to 1 %. These numbers reflect the changing power structures in Swedish society after the fall of absolutism and the rise of a strong parliament in the 1720s. This period saw changes in *who* or *what* the *publicum* constituted.[51]

The applicant's inclination to mining, his background within academia and in mining districts, as well as his wish to serve his fatherland by working in the Bureau of Mines, together formed a patriotic narrative of submission. One of the clearest examples can be found in Samuel Buschenfelt (the younger)'s application from 1725:

> I subserviently and humbly trust in the high favour and tendency that his lordship [i.e., the president of the Bureau] and the highly praised *Bureau of Mines* always choose to show those, who after *academic studies*, and after acquiring knowledge in mining, take the greatest pains to make themselves skilled in the service and use of their fatherland. Thus, I venture to request and implore of your lordship and the highly praised royal *Bureau* that [you] graciously would accept me as an *auscultator*.[52]

By studying the applications, we can thus observe patterns and developments in how auscultators presented themselves to the Bureau, in anticipation of the Bureau's criteria for identifying an apt young man. One could state that the route to legitimate peripheral participation in the Bureau of Mines was through the presentation of previous participation in relevant communities of practice, as well as the demonstration of skills and knowledge acquired there. The universities, especially Uppsala, were important references at the turn of the century, and had further increased their importance by the 1750s. During the study's time period, one can delineate a process of homogenisation of the applicant's background: especially in respects to the number of applicants who had conducted academic studies.

---

[49] Karin Sennefelt has also identified this connection between patriotism and the dismissal of private gain. See *Politikens hjärta*, 239–40.

[50] Widmalm: *Mellan kartan och verkligheten*, 18.

[51] Interestingly, in the first ten years of the century few applicants mentioned either serving the king of the patria. It seems as if the ideal of the civil service as a way of serving the king or the patria did not become widespread among applicants until some years into the eighteenth century.

[52] "I underdån ödmiuk förtröstan af den höga gunst och benägenhet Eders Grefl. *Excellence* och det Höglofl. Kongl. Bergz *Collegium* städze behagat wisa dem, som efter anlagde *studier* wid *Academien*, samt inhämtad kundskap i Bergssaker och handteringar, jämwäl på det högsta sig winlägga at till fäderneslandets tienst och nytta skickelige sinnat, fördristar iag för Eders Grefl. *Excell.* och det Höglofl Kongl. Kongl. [sic] *Collegium* jag underdån ödmiukast bedia och bönfalla, at af Eders Greflige *Excell*, och det Höglofl. Kongl. *Collegio* för *Auscultant* nådgunstigt blifwa antigen."; Samuel Buschenfelt: "Application to auscultate", 1725-11-08, E4/154/868, Bergskollegiums arkiv, RA.

But even at the end of the period, an academic degree was not the only way into the Bureau. Mentioning how one had lived in the mining districts during childhood or had visited them on one's travels was an additional way of presenting oneself as a participant in a relevant community of practice. Also, by discussing kinship, applicants could display connections to relevant networks and communities. In their applications, young men presented their life narrative in a way that corresponded to the same ideal childhood of a useful man, found in speeches and prefaces to textbooks. If we take the self-presentations of the auscultators as a sign of what skills were demanded by the Bureau, it seems as if by the 1730s it was important to present a proficiency in the mathematical sciences alongside experience of assaying and juridical matters. A young man who demonstrated aptitude for these skills was generally identified as a legitimate peripheral participant in the Bureau's community of men. Such a man became accepted into the group of unpaid *venia auscultandi*, or, if he was lucky, even as one of the paid *beneficium auscultandi*.

Mathematical sciences of the state

Among historians of science, the Bureau of Mines has primarily been discussed as a place of chemistry.[53] Still, in the applications from the first half the eighteenth century, few aspiring auscultators discussed chemistry explicitly. Instead, the craft of assaying was a reoccurring topic. At the beginning of the eighteenth century, chemistry was not a unified science. Instead, it consisted of various crafts, among them assaying. The lack of applicants discussing chemistry in their applications thus supports the claim that chemistry developed as an "art" from the 1720s to 1740s and a "systematic science" from the mid 1700s to the end of the century.[54] In comparison, in the applications of this time, mathematics was repeatedly used in a broad sense, as a unifying concept encompassing geometry, arithmetic but also so-called "practical mathematics". Although uncommon during the first decade, from 1700 to 1750, on average 18 % of the applicants mentioned mathematics, roughly comparable to the 16 % who mentioned assaying. While the roles of assaying and chemistry are clearly important in an enterprise that examines ores and metals to ascertain a high and consistent quality, mathematics and geometry are possibly less self-explanatory parts of such an institution.

---

[53] See Fors: *Mutual favours*; Lundgren: "Gruvor och kemi under 1700-talet i Sverige. Nytta och vetenskap"; Fors: *The limits of matter*. Also, as already discussed, Svante Lindqvist has discussed the Bureau as a place of mechanical knowledge making in his *Technology on trial*, esp. 95–107.

[54] Lissa Roberts: "Filling the space of possibilities. Eighteenth-century chemistry's transition from art to science", *Science in context* 6:02 (1993), 512, 548–9; Fors: "J. G. Wallerius and the laboratory of enlightenment", 17.

What role did the mathematical sciences have in the Bureau between 1700 and 1750?

As pointed out earlier, the Bureau was not the only part of the Swedish state where the mathematical sciences were practised. From the seventeenth century, it is possible to discern a social framework for the practice of mathematics in several institutions of the Swedish state.[55] Mathematical textbooks in the Swedish vernacular were generally produced in connection with the institutions of the civil administration and the military in Stockholm, and in these books we consequently find arguments for the use of mathematics in the civil service.[56] In 1614, Aegidius Matthiæ Aurelius presented the role of mathematics in the training of a civil servant. In his *Arithmetica* (first published in 1614, but printed in new editions well into the eighteenth century), Aurelius connected mathematics to good government. He argued that "no *politie* or any government, land or realm, town or village, not even the most simple hamlet, exists in the world, which does not need this art". Those who had "learnt this art, are much more comfortable, useful and skilled in all matters of the state, than those who are inexperienced in this art". Aurelius argued using classical examples: in classical Egypt, no one had taken up an office [*Åmbete*] "with less than that he first had practised and carried out [arithmetic] and other mathematical arts."[57]

---

[55] An important role of mathematics was in geodesy and map making. See Widmalm: *Mellan kartan och verkligheten.*

Larry Stewart describes English public science as an activity of improvement for the public good with wide implications within a rising bourgeois public sphere separated from the state; Stewart: *Rise of public science*, 38–40. See also Jan Golinski: *Science as public culture. Chemistry and Enlightenment in Britain, 1760–1820* (Cambridge, 1992), 4–8. For a discussion on eighteenth-century European technology as public culture which belonged "to political economy, to the sciences of organization and action", see Liliane Hilaire-Pérez: "Technology as a public culture in the eighteenth century. The artisan's legacy", *History of science* 45 (2007), 135 and passim. One way of understanding these sciences of public power is Pierre Bourdieu's concept of "la science de l'état" (i.e., "une science anonyme et pratique de l'administration et des fins et des moyens de l'État"); Pierre Bourdieu, Olivier Christin and Pierre-Etienne Will: "Sur la science de l'État", *Actes de la recherche en sciences sociales* 133:1 (2000), 5.

[56] Also, Sten Lindroth points out how a majority of members of the Swedish Royal Academy of Sciences during 1739–45 were civil servants at different levels within the Swedish state; Lindroth: *Kungl. Svenska vetenskapsakademiens historia 1739–1818 1:1*, 28.

[57] "ingen *Politie* eller någhot *Regemente*, Land eller Rijke, Stadh eller Byy, Ja, icke thet ringaste Torp i Werldenne finnes, thet thenne konst icke behöfwer"; "the som medh thenne konst begåfwadhe äro, warda til alle *Polisiske* sakar myket beqwemligare, brukeligare och skickeligare, än the som uthi thenne konst äre oförfarne"; "medh mindre han sigh tilförende uthi thenne och ander *Mathematische* konster öfwat och brukat hadhe."; Aegidius Matthiæ Aurelius: *Arithmetica eller een kort och eenfaldigh räknebook, uthi heele och brutne taal. Medh lustige och sköne exempel, them eenfaldighom som til thenne konst lust och behagh hafwe* (Uppsala, 1614), preface. On the biography of this early author of Swedish mathematical books, see B Boethius: "Aegidius Matthie Upsaliensis Aurelius", *SBL* 2 (Stockholm, 1920), 451. A similar argument was made in the early eighteenth century, in Eric Agner: *Geodæsia Suecana eller Örtuga delo-bok* (Stockholm, 1730), preface.

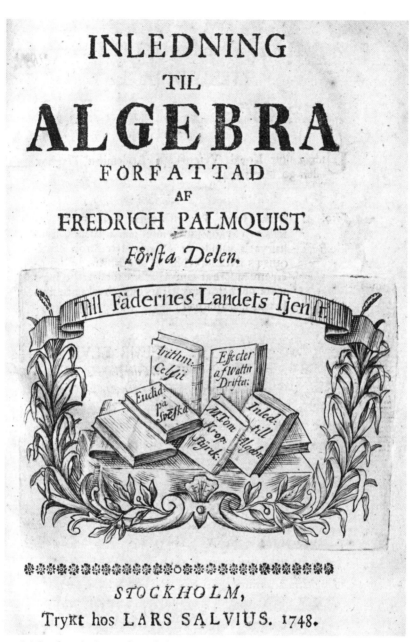

INLEDNING
TIL
ALGEBRA
FÖRFATTAD
AF
FREDRICH PALMQUIST
Första Delen.

Till Fädernes Landets Tjenste

Arithm: Celsii

Effecter af Watn Driften:

Euclid: på svenska

Inled: til Algebr:

M:Tom Krop: Styrk:

STOCKHOLM,
Trykt hos LARS SALVIUS. 1748.

Figure 4. The frontispiece of Fredrik Palmqvist's *Introduction to algebra* (1748). The text reads "In the service of the fatherland." Framed by a laurel are five mathematical works in Swedish, many of them discussed in the previous chapter. Besides Palmqvist's *Introduction to algebra* itself, the books were Celsius' *Arithmetica* (1727), Elvis's treatise on waterwheels *Mathematiskt tractat om effecter af vatn-drifter* (1742), Strömer's translation of Euclid's *Elementa* (1744), and Palmqvist's own treatise on *The solidity and strength of bodies* (1744). (Photo: Uppsala University Library)

During the seventeenth century, when the administration was less extensive and the auscultatory system was fledgling, books such as Aurelius' were relatively rare, but in the early eighteenth century there was a marked increase in the publication of mathematical textbooks, as discussed in the previous chapter.[58] The frontispiece of Fredrik Palmqvist's *Introduction to algebra* (1748) explicitly presented Swedish mathematical textbooks as patriotic works by placing a number of such books in a laurel under the celebratory text "in the service of the fatherland" (see Figure 4). By the mid-eighteenth century, vernacular mathematics would become a form of knowledge integrated with the state that would identify a man as a virtuous civil servant.

Mathematics as mechanics (1700–20)

As the number of vernacular publications increased, the roles of mathematics and mechanics changed in the Bureau of Mines. In an *extractum protocolli* of 28 November 1715, issued to all auscultators of the Bureau, the board argued that an auscultator was required to show that he had studied mining and smelting [*bergsväsendet*] and that he knew how to apply his acquired knowledge. Already at this time, the board stressed that "*mathesis*" was an important basis for a future official.[59] Correspondingly, between 1710 and 1750, it was relatively common to mention mathematics in one's application to the Bureau (see Figure 2 and 3). But what was meant by *mathesis* changed from the early to mid-eighteenth century: the group who mentioned mathematics in their applications expanded, as mathematics became a more prominent part of the mining sciences.

In the first two decades of the century, mathematics and mechanics were mostly discussed by applicants who aimed for the mechanical stipends of the *Laboratorium mechanicum*. The only applicant in the first decade who mentioned mechanics, Göran Wallerius, also explicitly mentioned Christopher Polhammar and the *stipendium mechanicum*. All the four applicants mentioning mathematics in the 1710s also mentioned mechanics. Three of them (Grave and Tilaeus, and Duhre) explicitly mentioned having studied with Polhammar, and three (Geisler, Grave and Duhre) applied for the stipend in mechanics at some stage. In the applications of the early 1700s, mathematics was thus almost exclusively associated with mining mechanics, and the dis-

---

[58] Like Aurelius' arithmetica, most books from the seventeenth century published for men of the civil service and the military were geared towards explicit uses, such as Peder Nilsson Raam's book on the geometrical measurements of fields from 1670 and the "Captein Ingenieur" Berthold Otto Smoll's introduction to geometry for officers of the fortification corps from 1692. Peder Nilsson Raam: *Then swenske åkermätningen, eller Ortuga deelo book, item een lijten tractat, om staaff och råå, och thes beskaffenheet* (Strengnäs, 1670); Schmoll: *Kort anledning till geometrien, hwar effter officerarne wid Hans Kongl. May.*

[59] "Protocol of the Bureau of Mines", 1715-11-28, A1/53/1817–24, Bergskollegiums arkiv, RA, 1819. For a more detailed discussion of this protocol, see page 110–11.

cussions of mathematics can be delimited to a clearly delineated subgroup of auscultators.[60]

These applicants' relationships with Polhammar is unsurprising, given his role at that time as the final authority on identifying mechanical competence in the Bureau. His authority meant that it was difficult for young men to assert their mechanical and mathematical skill without his support. Generally, a close and personal relationship with him was required to receive one of the mechanical stipends; in practice, it seems that what Polhammar considered to be the surest sign of mechanical competence was that you were his son, or at least one of his most valued students. In 1708, Magnus Lundström and a Mr Rangelius had applied for a mechanical stipend. In his response to the Bureau, Polhammar argued that he did not know what Lundström knew of *theoria* or *praxis*. As for Rangelius, Polhammar did not know anything of his studies in mechanics, but he knew that he was a skilled draughtsman. Therefore, he recommended Rangelius instead of Lundström. But, he continued, he especially recommended his son Christopher Polhammar the younger, who was on a European tour, and who, at the time, was staying in Leipzig. According to his father, Polhammar the younger was well versed in mechanics. The Bureau accepted Polhammar the elder's suggestion, and decided that his son should receive the stipend when he returned home and could present evidence of his mechanical skill, in the form of drawings and actual machines.[61]

Similarly, in 1715, when Duhre and Georg Rudin applied for the stipend, Polhammar instead recommended his younger son Gabriel. He considered Duhre to be competent in "*mathesi pura*" but argued that he was too old to learn the "mechanical *praxis*". Thus, again we see how distinctions of age were central to the perceived ability to learn mechanics. Instead, Polhammar argued, Duhre would "provide his Majesty a much greater service, if he were used as a teacher for the young [at the naval base] in Karlskrona in the things that belong to *navigation*." The board agreed that one of the stipends should be given to Gabriel Polhammar, but sidestepped the opinions of Polhammar Senior in the case of Duhre. Although Polhammar Senior obviously gave a negative review of Duhre's mechanical skill, the

---

[60] Göran Wallerius: "Application to auscultate", 1705-03-21, E4/116/206, Bergskollegiums arkiv, RA; Johan Tobias Geisler: "Application to auscultate", 1711-11-23, E4/129/518, Bergskollegiums arkiv, RA; Sebastian Grave: "Application to auscultate", 1712-11-24, E4/131/388, Bergskollegiums arkiv, RA; Petrus Tillaeus: "Application to auscultate", 1712-12-20, E4/132/2, Bergskollegiums arkiv, RA; Anders Gabriel Duhre: "Application to auscultate", 1715-09-26, E4/137/113, Bergskollegiums arkiv, RA. See also Figure 2 and 3.
[61] "Protocol of the Bureau of Mines", 1708-02-12, A1/46/130–42, Bergskollegiums arkiv, RA. Because Christopher Polhammar Junior died in Leipzig later that year, he never received the stipend. After Polhammar's son died, Magnus Lundström applied for the stipend again. See Magnus Lundström: "Application for stipendium mechanicum", 1708-12-02, E4/123/1004, Bergskollegiums arkiv, RA. Lundström did not receive the stipend at this time either. Instead, it was given to Harald Lybecker in 1709, see Almquist: *Bergskollegium och Bergslagsstaterna 1637–1857*, 113.

board responded as if he had been in favour of him. They wished to give Duhre the other mechanical stipend because of "the good recommendation from [Polhammar] [...] that he has put down an uncommon diligence in *mathesis*". The Bureau proposed that Polhammar should try to find Duhre a position in Karlskrona himself, and decided that they would give Duhre the stipend should he fail.[62]

Mathematics as the basis of a community (1720–50)

During the first two decades of the eighteenth century, the mathematics and mechanics of the Bureau were more or less a Polhammar family business. But the admittance of Duhre, against Polhammar's recommendation, could be seen as indicating that his authority was diminishing by the late 1710s. This is also the time when the number of applicants who mentioned mathematics grew. Gabriel Polhammar and Anders Gabriel Duhre kept their stipends into the 1730s. Consequently, in the 1720s, none of the growing number of applicants who mentioned mathematics applied for the *stipendium mechanicum*. Furthermore, in contrast to the applications of the first decades, none of these latter ones mentioned Christopher Polhammar. Lindqvist has pointed out that, by the 1720s, men such as Urban Hiärne and Christopher Polhammar were representatives of a past era, when men had been able to rise in the bureaucracy through royal command instead of through the "normal channels". At that time, having been a client of the former absolute monarch hindered rather than facilitated advancement.[63] To present a relationship with Polhammar was therefore not desirable at this time and the applicants of the 1720s must have presented mathematical skill for other reasons than to show their affiliation with him.

Also, by the 1730s, the strong correlation between mentions of mechanics and mathematics disappeared. In this decade, 19 (25 %) of the 76 applicants mentioned having studied mathematics and these studies' importance to mining work. Only six mentioned mechanics, and only one, Samuel Sohlberg, received the *stipendium mechanicum*. At this time, there was therefore a large group of young men who presented mathematical skill to the Bureau, but who did not necessarily associate this skill with mining mechanics. In the 1740s, 18 out of 88 applicants mentioned mathematics in their application, but only three mentioned mechanics. These applicants of the 1730s and 1740s did not necessarily just see mathematical exercise as a way to become a mining mechanicus. Instead, they discussed mathematics as

---

[62] "*mathesis pura*"; "*mechaniska praxin*"; "Protocol of the Bureau of Mines", 1716-04-23, A1/54/399–406, Bergskollegiums arkiv, RA, 400. "han skulle Kunna giöra Hans May:t en fast större tienst, om han blefwe brukad för en Läremästare för ungdommen i *Carlscrona* uti de saker, som til *navigationen* höra."; ibid., 400–401. "det goda loford och witnesbörd, af H. *Assessoren* och *Directeuren* är wordet lemnadt, at han uti *Mathesin* nu ogemen stor flit anlagt"; ibid., 404.
[63] Lindqvist: *Technology on trial*, 99.

relevant to the whole business of the Bureau. These young men recognised the Bureau as a community where the learning of mathematics was part of the process of becoming a full participant. The applicants, including Sohlberg in 1731, described how they had studied things "that facilitate mining, and the adherent mathematical and mechanical sciences." Similarly, in 1748, Gustaf Bierchenius wrote of how he, because of his interest in and inclination for mining, "especially had had endeavoured to learn *Mathesis* and *Physics*."[64] This function of mathematics post-Polhammar could be seen as a foundation of a shared social epistemology: it was a means of establishing a social framework based on a common way of seeing and thinking about the world.

This new role of mathematics is also found in the prefaces to two mathematical textbooks, written by auscultators of the Bureau in the late 1710s. In *A Fundamental Guide to* Mathesis Universalem *and Algebra* (1718), Brandt, accepted as an auscultator in 1715, argued to his superiors that mathematics was integral to the Bureau's work. Brandt's work, which was based on the lectures of his fellow auscultator Anders Gabriel Duhre, can be seen as an early forceful argument for the relevance of mathematics beyond mining mechanics. As argued by Fors, Brandt would later become the "grand old man" of the group of chemists who saw mechanical philosophy as a basis for their art.[65] But, in Brandt's preface, it is possible to discern an even wider role for mathematics: as a legitimate socio-epistemological foundation of all the Bureau's activities that would unite the officials through a common framework of seeing and thinking.[66]

Brandt's publication was dedicated to the Bureau's vice president, the mine councillors [*Bergsråd*] and the Bureau's assessors. In his preface, Brandt firmly placed himself and his publication within the hierarchies of the Bureau. Written in this context, Brandt's book on algebra should not be considered a scholarly text. It was written in the vernacular instead of Latin, the royal printer in Stockholm (Johan Horrn) printed it rather than the printer in Uppsala, and it was meant for Brandt's fellow civil servants, rather than for a European scholarly audience. It can thus be seen as an example of the tradition of mathematical textbooks in Swedish produced in the bureaus in Stockholm, discussed earlier. From the start, Brandt conformed to the style of writing expected of a Stockholm cameralist. He humbly pointed out that his book should not be seen as a work "of any perfection". His only intention was to give his "gracious superior" a "simple specimen of the diligence" that he had put into the mathematical sciences while being an

---

[64] "som till BergzWärks, och de der til hörande *Mathematiske* och *Mechaniske* Wettenskaper tiäna."; Samuel Sohlberg: "Application to auscultate", 1731-05-25, E4/167/946, Bergskollegiums arkiv, RA. "i synnerhet winlagt mig om *Mathesin* och *Physiquen*."; Gustaf Bierchenius: "Application to auscultate", 1748-02-29, E4/217/42, Bergskollegiums arkiv, RA.

[65] Fors: *The limits of matter*, 95.

[66] On the relation between mathematics and chemistry in Brandt's preface, see ibid., 91.

auscultator. The mathematical sciences, he continued, were the foundation of "as many parts of the correct practice of mining, as of much other highly useful knowledge".[67]

Brandt's preface put forth an epistemic framework in which mathematical knowledge, work and production were intertwined.[68] An important aim of the preface was to establish mathematics as a foundation for several branches of the Bureau's work. Through mathematics, mining could "be noticeably improved".[69] The preface argued that:

> What great service and usefulness that the *Mathematical* sciences provide for the mines is such a well-known matter, especially what concerns *Mechanics,* that there is no need to describe it further.[70]

Thus, Brandt's aim was not to link mathematics to mechanics, nor to contest that mathematics was relevant to this field. The connection between mathematics and mining mechanics had already been done by Polhammar and, according to Brandt, was self-evident for the Bureau's officials. For Brandt, it was instead central to show that by applying mathematics to the experience of nature it was possible to establish physical facts. The prime example of such a practice, according to him, was "the splendid *mathematical* and *philosophical* works by the Englishman *Newton,* which clearly show how much in physics that can be discovered, using the help of mathematics."[71]

Anders Gabriel Duhre, discussed earlier and whose lectures Brandt's publication was based on, was a mathematician with an interest in mechanical inventions. Like the other officials of the Bureau, and following Polhammar, Duhre considered mathematics, and more specifically geometry, to be the foundation of mechanical work. But in 1722, when arguing for an educational institution for the teaching of œconomical sciences using mathematical principles, Duhre, similarly to Brandt, argued for a broader role for mathematics. He pointed out how "Chemical *praxis* and the han-

---

[67] "af någon fullkomlighet"; "min nådigste förman"; "ett ringa prof af den flijt"; "många delar till Bergwärks rätta handhafwande än andra högnödige kundskaper sig grunda"; Brandt: *Mathesin universalem,* dedication.

[68] Ibid., title page. Other works, discussing a similar role of mathematical and physical sciences in the Bureau are for example Duhre: *Förklaring,* 23–7; Ekström: *Tal, om järn-förädlingens nytta och vårdande;* Göran Wallerius: *Tal emellan mathesin och physiquen om deras verkan och nytta uti bärgs-väsendet, förestäldt i kongl. svenska vetenskaps academien af Göran Vallerius vid præsidii afläggande den 21 jan. 1744* (Stockholm, 1747); Isaac Johan Uhr: *En brukspatrons egenskaper* (Uppsala, 1750).

[69] "märkeligen befrämjas"; Brandt: *Mathesin universalem,* dedication.

[70] "Hwad stor tienst och gagn de *Mathemati*ske wettenskaper tilfoga bergwärken, är hos oss i synnerhet hwad *Mechaniquen* widkommer en så bekant sak, at der om intet behöfwes wijdare förmåla"; ibid., preface.

[71] "den widt berömde Engels-Mannen *Newtons* förträffeliga *Mathemati*ska och *Philosophi*ska wärk, som klarligen wijsa huru mycket medelst *mathematiquens* tilhjelp, der sammastädes uti *physiquens* uptäcktes som tillförende warit omöyeligit innan *application* af *Mathesi* blef kunnig."; ibid.

dling of *metals*", would be "improved through the application of *geometry*."[72] Whereas Polhem had been the director of the *Laboratorium mechanicum* in the Bureau, Duhre proposed that he should be the director of a *Laboratorium mathematico-œconomicum*, signalling that mathematics was the basis for all useful œconomical knowledge. *A Fundamental Guide* might be seen as the joint product of the mathematically inclined men Brandt and Duhre, both inspired by Newtonian science and eclectic philosophy in their views on the role of mathematics in material and knowledge production. Duhre and Brandt presented the Bureau as a community that was (or at least should be) formed around a natural philosophy based on mathematical principles. This community should act according to common and predictable methods, knowledge and norms, based on mathematics.

As seen from the protocols of the board from 1715, as well as from the applications to auscultate from the 1710s onward, Brandt's linking of mathematics and mining in 1718 was hardly original. Still, *what* the unifying term of "mathematics" contained and *how* its contents were related to mining was not static. We could thus see the statements of Brandt and Duhre as arguments for a specific epistemology that gave mathematics a new and prominent position in the Bureau. What the two argued for in their publications was the reconfiguration of the category of the public man into one who based his work on an experimentalist framework underpinned by mathematics and especially geometry. In the 1720s, out of the sixty-one young men who applied to the Bureau three mentioned mathematics, mechanics *and* physics in their application. One of these was Daniel Bergenstierna, who took private classes in mathematics from Duhre. In his application in 1723, Bergenstierna explicitly referenced Duhre as his teacher, and more or less repeated the arguments for geometry found in Duhre's publication from the same year. He stated that he had studied the parts of mathematics, "namely *geometry* and *algebra*, which provide a good basis for physics and mechanics". He had acquired such mathematical skills by having taken "advantage of Mr Anders Duhre's faithful teaching with the aim, that I more easily can acquire the mining science".[73] By the 1740s, many applications were similar to Bergenstierna's in that they linked *mathesis* to *physica* and to the mining sciences.

---

[72] "*Chymiska Praxi* och *Metallernas* handterande […] böra befrämias igenom *Geometriens Application*."; Duhre: *Förklaring*, 26. For further discussion of Anders Gabriel Duhre and his educational institution, see Chapter 5.

[73] "nämligen *Geometrien* och *Algebra*, som gifwa anledning till en god grund uti *Physiquen* och *Mechaniquen*"; "deruthennan någon tid borråt nyttiat Hr *Anders Duhres* trogna underwisning till den ändan, att jag sedan destebättre Bergwärks kundskapen måtte kunna inhämta"; Daniel Bergenstierna: "Application to auscultate", 1723-10-24, E4/151/307, Bergskollegiums arkiv, RA. The other applicants discussing mathematics were Samuel Buschenfelt the younger, the son of Christopher Polhammar's travel companion (1725), and Samuel Schultze (1726).

The mathematical man of metals

Which demands did these young men anticipate, who discussed mathematics in their applications? The role of mathematics in the Bureau of the mid 1700s can be found in the Uppsala dissertation *The characteristics of an ironmaster* [*brukspatron*, literally "patron of a *bruk*"] written in 1750, which describes this mathematical man of metals. The thesis was presented by the auscultator of the Bureau, Isaac Johan Uhr, under the *praeses* Anders Berch, the professor of œconomy in Uppsala. In his dissertation, Uhr discussed what character, knowledge and skills an owner and director of an ironworks should possess, and argued for the importance of mathematics and physics in the making of such a man. Although the ironmasters were not the same men as the civil servants within the Bureau (however, many auscultators came from families of ironmasters and many auscultators eventually became ironmasters), it is clear that Uhr's beliefs as to the knowledge and skills an ironmaster should possess reflected the norms of the community that he was becoming part of. As such, his discussion on the role of mathematics in becoming a man of mining and smelting is informative of the norms for the role of mathematics in the Bureau of the 1730s.[74]

Uhr began his work with the common metaphor of society as a body consisting of diverse parts. As a part of this body politic, the role of the ironwork's owner, much like that of civil servants, was to act in the interest of the whole of the state. Because the ironmaster was an important member of the social body, it was vital that he was knowledgeable in his trade and that he constantly strived to perfect it. According to Uhr, it was because of the ironmaster's desire for continuous improvement, in the interest of the state, that metal making and mathematics merged. Such practices of improvement not only improved his ironworks: they were also exercises by which he bettered himself. Like the writers discussed in the previous chapter, Uhr presented mathematics as part of a process of becoming; he explicitly pointed out that natural knowledge, good quality metal and virtuous men were created in parallel:

> Thus, when [the ironmaster] examines ores, and wishes to perfect the knowledge about them, he also examines and perfects himself to quite a useful member of society, and himself becomes both the worker in, and the product of, his workshop.[75]

---

[74] Uhr: *En brukspatrons egenskaper*. It is generally difficult to know if an eighteenth-century academic dissertations was written by the professor, the præses or the respondendt. However, according to Sven-Eric Liedman, because of how the style of this dissertation diverged from the others under Berch, it is likely that Uhr wrote it himself; Sven-Eric Liedman: *Den synliga handen. Anders Berch och ekonomiämnena vid 1700-talets svenska universitet* (Stockholm, 1986), 119.

[75] "Således, då han pröfvar Malmer och vill upbruka kunskapen om dem, pröfvar och upbrukar han sig sjelf til en ganska nyttig lem uti Samhället, och blifver sjelf både verkande och verkstad uti sin verkstad."; Uhr: *En brukspatrons egenskaper*, 25.

Such examination of nature, which made an ironmaster into someone other than a commoner working with mining and smelting [*bergsman från allmogen*], was based on mathematical and physical principles. Using the common trope of the difference between simple craftsmen and the *mechanicus*, Uhr argued that common men, who only had the bodily skills required for menial labour, were merely able to carry out "what cultivated minds had conceived."[76] However, men whose minds had been cultivated by the mathematical sciences had an obligation to "to cultivate their natural senses to advance the public good".[77] The ironmaster, like the *mechanicus*, was defined by his cultivated and imaginative mind that could invent new things. Such men should have as their objective to "perfect their knowledge of the subject [of mining], and to constantly work to carry the arts to greater heights and perfection."[78] For Uhr, the repetitious practice of certain skills and gestures, which improved both knowledge and the knower, was what set the ironmaster apart from other men. He specified how "the skill of an ironmaster begins with knowledge in *mathematics, physics* and *chemistry*".[79] Furthermore, an important part of mathematics was mechanics, which taught the ironmaster "the correct and suitable construction of buildings."[80] These skills identified an ironmaster as "a father, who gave birth to and nurtured as many useful members [of the state], as the number of young men he made into useful workers."[81] Thus, again, mathematics and mechanics were imagined to be part of a process of maturity, and a means of asserting fatherly authority. By engaging with practical mathematics, a man of mining and smelting would mature into a patriarch worthy of ruling other subjects of early modern Sweden.

In Uhr's narrative, mathematical and natural philosophy constituted a boundary between the ironmaster and other men found in metal production. Likewise, by the 1730s, the officials of the Bureau identified themselves as socially and epistemically distinct by the manners in which they related mathematical and physical knowledge to their work. Mathematical skill came to carry connotations similar to those of a university degree, as seen, for example, in the quote from Fahlander's application of 1750, discussed earlier (see page 89–91). Like a degree, mathematical proficiency was not only evidence of a set of skills, but also proved that you were a specific kind of man, who was separated from the common crowd and suited for work in the Bureau. As the auscultatory system grew, mathematics became

---

[76] "som upbrukade snillen uptänkt"; ibid., 17.

[77] "at förädla sina naturliga sinne gåfvor til befrämjande af det almänna bästa"; ibid.

[78] "at göra sig sjelfva fullkomliga i saken, och ständigt arbeta på konstens upbringande til större högd och fullkomlighet."; ibid.

[79] "Börjar altså en Bruks-Patrons skicklighet i kunskapen uti *Mathematik, Physik* och *Chemie*"; Ekström: *Tal, om järn-förädlingens nytta och vårdande*, 18–9.

[80] "bygnaders rätta och riktiga inrättning"; Uhr: *En brukspatrons egenskaper*, 19.

[81] "en Far, som födde och upfödde så många och nyttiga lemmar, som han gjorde sådane ynglingar til gagneliga arbetare."; ibid., 27.

an important signifier by which the board could identify legitimate peripheral participants. In the applications to auscultate from the 1730s and 1740s, we can see the rise of a large group of young men who wished to adhere and align themselves to the persona of a mathematical civil servant in order to participate in the Bureau's work. Mentioning mathematics in one's application thus held an important symbolic function, which by the 1730s was not only related to such activities as machine building or subterranean construction work.

Becoming a Bureau official

The processes of self-formation and identification did not stop after the board had identified an apt auscultator. Quite the opposite: although auscultators already conformed to anticipated demands in their applications, once accepted they were formed by a process of enculturation that ideally resulted in a permanent position as an official. Within a couple of days of admittance, the auscultator made both an oral and a written oath. The oral oath was made in front of the board and registered in the protocol; the written oath was kept in the Bureau's archive. In his history of the Bureau, Daniel Tilas attached a standard form for the auscultator's oath. The auscultator should swear to work diligently and, with respect and reverence, he should carry out what his superiors wished him to do. He should also make a pledge to God not to share any of the matters discussed by the board with outsiders.[82] The auscultators' written oaths found in the archive follow this standard document to a large degree, and they were thus generally written in a standard form. The written oath was not a presentation of who the auscultator was, but of what he would become; it was a promise that he would conform to the norms of the Bureau and that he would shape himself in the image of a virtuous and submissive civil servant. Therefore, it is only possible to discern some long-term trends in this otherwise homogeneous material, such as the fact that by the 1710s the auscultators had stopped pledging their loyalty to the king and instead swore allegiance only to the patria.[83]

In his commemorative speech over the former auscultator Lars Benzelstierna, Olof Celsius described Benzelstierna's acceptance as an auscultator as if "here, a new academy opened for this young man, in which he, from the most skilled masters, could learn the application of the theoretical knowledge that he had acquired in the previous [academy]."[84] Others were

---

[82] Daniel Tilas: "Om nuvarande Bergskollegium", 269.
[83] "Auskultantmatrikel", 1678-1815, D1/15, Bergskollegiums arkiv, RA. This development thus interestingly corresponds to the changed ways of discussing the *publicum* in the applications, as discussed earlier.
[84] "En ny Academie öpnades här för vår yngling, uti hvilken Han af de skickeligaste mästare fick lära tillämpningen af den Theoretiske kunskap, hvilken Han vid den förra hade inhämtat."; Olof

more critical of the Bureau's education. In a speech about Axel Fredrik Cronstedt, Sven Rinman described how "at first, our young man of mining and smelting must have been somewhat perplexed, when he was ordered to make a fair copy of the Bureau's letters and judgements". Cronstedt had wished to acquire new knowledge in the Bureau, but at first had not known how to learn anything from the tasks that were set before him.[85] In his *Curriculum vitæ*, Daniel Tilas was even more dismissive of his first time as an auscultator. During the autumn of 1732, he had stayed in Stockholm and had spent his time "more walking around in the streets, than doing anything really useful, because I, and all the other auscultators, lacked instruction and direction to useful work." Only after having contacted his uncle and neighbour in Stockholm – the vice-clerk of the Bureau, Erland Fredrich Hiärne – did Tilas' luck improve. Hiärne advised him to diligently collect and copy travel reports, and to start gathering specimens for an ore and mineral cabinet.[86]

The auscultators were not alone in having a negative view of auscultation: the satirists in Stockholm also picked up on the theme. For example, the weekly periodical *The Swedish Argus* described how young men became auscultators in order to live lives of drunkenness and debauchery:

> The kid enters the coffeehouse or the tavern, so self-important as if he wished to say: Hi, here I come! In there, the pipe, the cup, and the nip or the digestive are his exercises. One does not hear many wise words from his mouth. When I ask around, why he is left alone so carelessly? Someone answers: He is *employed* by a *bureau*. Then I don't dare say a word. Everyday when his father leaves for his vocation, the boy goes out and does whatever he pleases. He is an *auscultator*; he wanders around, hears what *happens* in town and passes judgement, so that old people must be silent and feel ashamed. Is that not to spoil the young?[87]

Celsius: *Åtminnelse-tal öfver lands-höfdingen, bärgs-rådet Lars Benzelstierna, hållit för Kongl. vetenskapsacademien den 6 dec. 1758* (Stockholm, 1759), 14. Similarly, Anders Berch talked of the Bureau as a university [*Hög-schola*] in his speech over Henric Kalmeter; Berch: *Åminnelse-tal öfver Henric Kalmeter*, 8.

[85] "I början blef väl vår unge Bergsman något förvirrad, då Honom förelades at renskrifva Kongl. Collegii Bref och Domar"; Sven Rinman: *Åminnelse-tal öfver framledne bergmästaren ... Axel Fred. Cronstedt, på kongl. academiens vägnar, hållet i stora riddarhus-salen den 6. martii, 1766* (Stockholm, 1766), 12.

[86] "mera at släntra gatorne omkring, än at giöra mig något verckeligt gagn, emedan manuduction och anvisning til nyttige arbeten feltes så väl mig som alle andre auscultanter."; Daniel Tilas: *Curriculum vitæ I–II, 1712–1757 samt fragment av dagbok september–oktober 1767*, Holger Wichman (ed) (Stockholm, 1966 [1712–57]), 33.

[87] "Ungen komma in på Caffé-Huset eller Källaren, så dryg, som han wille säija: Hij här kommer jag! Der äro pipan, tassen, och supen eller afsättaren, hans öfningar. Icke många kloka ord får man der höra af hans munn. När jag frågar effter, hwi han släppes så handlöst? Swarar man: Han är *engagerad* i ett *Collegio*. Då törs jag icke säija ett ord. När Fadren går om dagen til sitt kall, så går Poiken ut och giör hwad han will, han är en *Auscultant*, han wandrar omkring och hörer hwad som *passerar* i Staden och fäller omdömen deröfwer, så at gammalt folk måste tiga och

For its eighteenth-century readers, the strength of the satire must have lain in the contrast between its presentation of a young auscultator and the expected self-presentations of actual young men of the bureaus. According to the *Argus*, the auscultators were too young to work in the Bureau. They were not men, but boys, and therefore they could not be trusted to carry out their duties. Together with the auscultators' own observations, this piece from the *Argus* reminds us that the meanings ascribed to auscultation, and the role of mathematics in this education, should be seen as meaningful in relation to the community of men of the Bureau. For actors outside the Bureau, or when the men of the Bureau presented themselves to other audiences, auscultation could carry completely different meanings.

While there are diverse accounts of how diligent the auscultators actually were, hard work and studies still constituted a norm, repeatedly expressed by both the board and the auscultators themselves. On 28 November 1715, the members of the board treated the applications of two young men: Anders Bratt and the already mentioned Georg Brandt. The board considered both men worthy of auscultation: they had produced a suitable submission, shown an understanding of assaying and the domestic metal production, and expressed an interest in travelling abroad to learn more. Having quickly accepted the two men, the board also decided to communicate to them, as well as the rest of the auscultators of the Bureau, what was expected of them *"per Extractum protocolli"*. In the protocol, which I mentioned briefly earlier, the board put forth what an auscultator should learn in order to advance in the Bureau. The board communicated that the Bureau's positions were given according to the auscultators' "personal capacity", and no heed was to be taken of the amount of time he had served. Thus, the identification of skill should not end with the application: auscultation itself was a forum for presenting oneself as a capable man and this was done through continuous diligent practice. In this way, the auscultator should make evident that he was "not only to be familiar with *mathesis*, he should also make himself informed in all the pieces that are necessary to know when mining". He should know how to identify minerals, assay the quality of the ore and know how to mine it; he should know how to produce coal, and be knowledgeable of the processes involved in producing metals. These skills could be acquired by visiting domestic mines and ironworks and then by learning more on a foreign tour. From these voyages the auscultator should write reports of what he saw and submit them to the archive.[88] Thus, the Bureau was very clear in its expectations of accepted auscultators.

---

skiämmas. Är icke detta at skiämma bort ungdomen?"; Olof von Dalin: *Then swänska Argus* vol. 5 (Stockholm, 1732), [6].

[88] "personernes *capacitet*"; "bör icke allenast wara grundad uti *Mathesi*, utan ock på det nogaste giöra sig underrättad uti alla de stycken, som til Bergzwerckshandteringen äro nödige at weta"; "Protocol of the Bureau of Mines, 1715-11-28", 1819. This protocol seems partly to contradict Almqvist, who argues that no special competence was required for admission until the mid-

The already discussed "Short report" suggested similar ways to perform diligence within the Bureau. Once accepted, an auscultator should visit the Bureau's archive, where he should access and transcribe documents related to the mining œconomy. He should also visit the assayer 1–2 hours every day, and when the clerk was away the auscultator should offer to write protocols. After 1–1.5 years, the auscultator should apply for a permit [*promotorial*] to travel to the mining districts, where he could observe the mines and the smelting process. The "Short report" recommended a specific route of sites that the auscultator should visit, and gave examples of the various forms of useful knowledge that could be acquired at these locations. The places should be visited in a specific order, forming an itinerary in the area between the north shore of Lake Mälaren and the mining town of Falun. The auscultator should also read reports from previous officials who had travelled, both domestically and abroad. Ideally, in the end, the auscultator should apply for funds and permission to travel abroad, preferably to Hanover, Saxony or Bohemia. By doing these things, and by showing "respect and obedience," the young man was promised secure prospects in the mining administration.[89]

The everyday work of an auscultator thus involved numerous and diverse practices, ranging from domestic and foreign voyages to the copying of protocols found in the archive. In order to understand the whole process of becoming an official, one would therefore need to examine all these ways to become enculturated into the administrative community of practice. But here I have somewhat different focus: as stated earlier, I study the role of mathematics (in the early modern broad sense) in these processes, in order to understand the ways in which mathematical exercises formed an early modern subject. In order to understand mathematics as an exercise of self-formation in the Bureau, and especially with regard to geometry and mechanics, one little-studied category of sources is more relevant than the others: manuscripts on mathematics, mechanics and natural philosophy that the auscultators copied and from which they learnt to see the world in a new way.

---

eighteenth century, Almquist: *Bergskollegium och Bergslagsstaterna 1637–1857*, 46. Also, the protocol stated that a capable auscultator should be knowledgeable of the law, in order to be able to act as a judge and an experienced man of the metal production. On the role of foreign travels in the education of the Bureau officials, see Lindqvist: *Technology on trial*, 121–7; Göran Rydén: "The Enlightenment in practice. Swedish travellers and knowledge about the metal trades", *Sjuttonhundratal* (2013), 64–9; Fors: *The limits of matter*, 8, 53–8.

[89] "*Respect* och lydno,"; Schultze: "Kort betänkande huru en yngling som tänker söka sin fortkomst vid bergsväsendet bör sin tid anlägga", unpaginated.

Transcription as enculturation

The "Short report" repeatedly pointed out the importance of copying key texts found in the Bureau's archive: auscultators should ask the "recording clerk" for access to the "mining ordinances, as well as other letters and resolutions, which concern the mining œconomy of the realm". These texts should be transcribed in full, or at least copied in extract.[90] Because these texts were "numerous and extensive," the young men should be "diligent in this endeavour, so that not too much work is wasted."[91]

In two separate collections of the library *Carolina Rediviva* at Uppsala lie two volumes filled with the products of the auscultators' work. Brandt, encountered earlier, signed the first of these two collections and he most likely either copied down or authored these texts when he auscultated in the Bureau. His collection contains texts on geometry and mechanics and their role in mining, as well as a treatise on mechanistic natural philosophy.[92] The second collection of texts contains some of the same texts on geometry and mechanics found in Brandt's manuscript, but lacks many others. This collection is most likely written in the hand of another auscultator: Lars Schultze.[93]

In Schultze's collection of manuscripts, one also finds the "Short report", discussed earlier. Thus it is likely that this report was a product of the very practice of copying that it prescribed.[94] The possibility that neither Schultze nor Brandt was the originator of these works does not make them uninteresting. Quite the opposite: it is because the handwritten texts were copies that they are relevant to this study. Copying was a technique for cir-

---

[90] "*actuarien*"; "Bergz fårordningen, och andre till Bergs *Oecononomien* här i riket hörande bref och *Resolutioner*", ibid.

[91] "många och wid löftiga,"; "wara så mycket flitig man på det intet, at för lång tid dertill må upgås."; Schultze: "Kort betänkande huru en yngling som tänker söka sin fortkomst vid bergsväsendet bör sin tid anlägga".

[92] The collection of Brandt's handwritings contains, among others, the texts Georg Brandt: "Om geometrien", A 28, 41–81, UUB; Georg Brandt: "Kort handledning till Driftkonsten eller Mechaniken", A 28, 245–302, UUB; Georg Brandt: "[On Philosophia naturali]", A 28, 334–465, UUB; Georg Brandt: "Om perspektivkonsten", A 28, 538–67, UUB.

[93] Sten Lindroth: *Gruvbrytning och kopparhantering vid Stora Kopparberget intill 1800-talets början. 2, Kopparhanteringen* (Uppsala, 1955), 16, note 8. I agree in Lindroth's assertion that the handwriting of e.g. Schultze's first application to auscultate, or his request for a dometic study tour, is the same as that found in this manuscript; see e.g., Lars Schultze: "Application to auscultate", 1726-11-22, E4/159/920, Bergskollegiums arkiv, RA; Lars Schultze: "Request for domestic study tour", 1729-02-12, E4/162/263, Bergskollegiums arkiv, RA. This second collection of texts contain Schultze: "Kort betänkande huru en yngling som tänker söka sin fortkomst vid bergsväsendet bör sin tid anlägga"; Lars Schultze: "Om geometrien", C 1433, UUB; Lars Schultze: "Kort handledning till Driftkonsten eller Mechaniken", D 1433, UUB.

[94] It has been suggested that Anton Swab could have authored the "report", as Schultze copied other texts by him and because Swab's education corresponds better to the career path that the guide suggests. Staffan Högberg: "Inledning", *Anton von Swabs berättelse om Avesta kronobruk 1723* (Stockholm, 1983), 15. But as Schultze arguably copied texts by a number of authors, whose education better fit the career path suggested in the guide, it is hard to pinpoint an original author of this text.

culating educational texts among officials: by spending time in the archive, the auscultator could reproduce authoritative knowledge and take it with him when leaving Stockholm. Thus the archive functioned as a collective memory for the community, and also as a central node that coordinated the communication and actions of its officials. The "Short report" in Schultze's collection pointed out that, although copying and excerpting texts was time-consuming work that should be done effectively, it was worth the effort. The Bureau was a community of practice, and these copying practices were part of a process of becoming a full participant in this community. By transcribing documents, an auscultator would internalise the norms that they expressed, in order to be enculturated into the Bureau and thus become an official. Also, handwritten texts held a value of exclusivity, compared with printed texts. Not everyone could copy such manuscripts: only those deemed worthy to be legitimate peripheral participants. Thus, possessing a library of your own handwritten texts, directly copied from the archive, was a sign of membership, as well as a way to possess a sample of the Bureau's collective memory.

The act of transcribing did not only reproduce the archive as new handwritten copies, but should also be seen as a means, as it were, to colonise the auscultators' minds with the methods and epistemologies of the Bureau.[95] Although the handwritten texts were not the original works of the auscultators who wrote them, and thus do not necessarily say anything about the opinions of their authors, we can infer from them the processes of becoming a member, and adopting the views, of the community. By making excerpts and transcriptions, auscultators did not only make copies of texts. This act of copying was also a means by which the auscultators made themselves into men who embraced a certain mathematical and mechanical way of seeing that was shared by the Bureau's officials. Among the handwritings found in Brandt's and Schultze's collections, four texts are especially informative of the roles of mathematics, mechanics and mechanistic philosophy as foundations of a community of men in the Bureau: first, "About geometry", found in both collections; second, "On the art of perspective", found only in Brandt's collection and which discusses seeing and drawing using geometric principles; third, the compendiums "A short instruction to the art of operation [driftskonsten] or mechanics", also found in both collections; and finally, a text on the "Knowledge of nature, or physics" found only in Brandt's collection. Together, these texts express a social epistemology of geometric perception and mechanical work. By studying them, we can understand the symbolic role of mathematics – in a broad sense – in forging a community of men in the Bureau of Mines after 1720.

---

[95] Compare with Shapin's discussion of the great civility as "granting the conditions in which others can colonize our minds and expecting the conditions which allow us to colonize theirs"; Shapin: *A social history of truth*, 36.

Geometric perception

The handwritings on geometry were placed first among the mathematical texts of both collections. They were a starting point and a basis for the other documents, which related to geometry as the foundation of their respective subject matter. This was also how the manuscript treated its subject matter: first it defined an ideal geometric point, and how it differed from a physical point in space, then straight and curved lines, and eventually surfaces and bodies. After the basics came some methods and problems to do with measuring and transforming geometric objects. The geometric point became a stable point of origin, from which the manuscript could continue with other definitions and methods of increasing complexity.[96] "About geometry" explained that geometry meant "the measurement of the earth", which was an unfortunate term because "she does not cover only the earth, but also all other magnitudes – consisting of lines, figures or bodies – in the *universe.*"[97] Geometry was thus both a structuring principle inherent in nature and a method to perceive these natural structures. By copying this manuscript, the auscultator would internalise a mathematical view of nature shared by the Bureau's officials, and also learn mathematical methods suitable for discerning this geometrically ordered world.

The text "On perspective" discussed geometry as a framework of perception more closely. Placed later in the collection, the auscultator would be likely to copy this text after he had studied and transcribed "On geometry". The text explained how perspective drawing was a technique of viewing the world through "a window or another glass" while "keeping the eye still". A perspective draughtsman would view the world from a clearly defined vantage point, from which one might see "a church, building etc." Using "a brush or a pencil", it was possible to "mark the *points* or *lines*, which show themselves there". Although such drawing on glass only required "a sure eye, and skill", it was still troublesome. Therefore, "*architects, mechanici* and painters use the *camera obscura* that, through a *convex* lens, *represents* on a paper everything that is outside." In some cases, however, none of these methods were applicable. Then the auscultator needed to use "geometric instruments". It was therefore important for a perspective draughtsman to know "the rules, which *mathesis* provides".[98] Perspective drawing, as presented in

---

[96] Thus, the structure of the text is similar to contemporary printed textbooks in geometry and arithmetic, published by members of the Bureau. See e.g., Brandt: *Mathesin universalem*; Duhre: *Geometria*; Weidler and Mört (tran.): *En klar och tydelig genstig.*

[97] "jordmätning"; "hon sträcker sig intet allenast till jorden, utan ock till alla andra *magnitudines*, bestående i linier, *figurer* el:r kroppar, uti hela *universe.*"; Brandt: "Om geometrien", 41.

[98] "ett fönster eller annat glas"; "håller ögat stilla"; "någon kyrka, bygning etc."; "en pensell eller penna"; "de *Puncter* och *linier* som der sig wisa"; Brandt: "Om perspektivkonsten", 538. "ögnamåttet och handtaget"; "så betiena sig *Architecter, Mechanici* och måhlare af *Camera obscura*, som på ett papper igenom ett *convext* glaas, *representerar* alt hwad derutan före är."; ibid., 539. "*Geometri*ska *instrumenter*"; "de reglor, som *Mathesis* gifwer thertill wid handen"; ibid., 540.

the manuscript, was thus a technique of perceiving the geometric universe discussed in "On geometry".

None of these manuscripts contained advanced or original mathematics. For a university graduate, and hence for most auscultators, their geometric definitions should have been well known. But originality was not their purpose; instead, these manuscripts presented the Bureau's framework of perception. Geometry established a way to perceive nature as simple parts – lines, figures and shapes – which could be easily described and communicated between the members of the Bureau. As discussed in Chapter 2, in early modern Europe mathematics was commonly imagined to be a technique of discernment and consent. Through mathematical exercises, a young man would not only learn to see a hidden truth in God's creation, he would also align himself to the theocratic order of the Swedish state. Also, as shown earlier, textbooks routinely presented mathematics as a patriotic form of knowledge, and as intimately linked to the state administration. In the handwritings by Schultze and Brandt, these two roles of mathematics converged. For the officials, geometry became a foundation for various techniques of perception, which could be used to visualise mines (as done through the mathematical art of *architecturam subterraneum*) and to make sense of machinery or other objects (through *mechanica*). But, as discussed in Chapter 2, mathematics (and especially geometry) was also imagined to foster a perceptive man aligned with the political order. In the Bureau, this imagined role of mechanics became the foundation of a geometric community of men. Similar to how a boy was imagined to attain a manly authority through mathematical exercise, from the 1720s, geometry became the basis for a male community of civil servants who policed the realm in the interest of the *publicum*. When the young auscultator sat down to copy the manuscripts on geometry, he also adopted some of the norms of this geometric community.

The mechanical powers

Whereas the texts on geometry and perspective drawing taught the auscultator to perceive a measurable world in a suspended state, the "Short introduction to [...] mechanics" unsurprisingly introduced the concepts of motion and power. Thus, after having transcribed the basis for perceiving the world geometrically, the auscultators turned to a framework of how to perceive motion in this geometric universe, as well as how to manipulate its regular lines and figures. Both versions of the manuscript pointed out how, with geometry as a basis, one could now "continue with the fundamentals of mechanics."[99]

---

[99] "fortfara uti mechaniska Grund-Satserne."; Brandt: "Kort handledning till Driftkonsten eller Mechaniken", 246; Schultze: "Kort handledning till Driftkonsten eller Mechaniken". Where

The introduction to mechanics commenced in equilibrium, before moving on to the "powers of mechanics". Thus, the first section of both manuscripts began by discussing *centro gravitatis*, or the centre of weight of bodies, how the weight was identical on all sides of this centre, and how the centre of bodies in free fall followed a straight geometric line towards the centre of the earth.[100] From copying down these basics of statics, the auscultators learnt that the world was a predictable place, and that matter followed certain prescribed geometric paths. Thus, for the men of the Bureau, even a world in motion was a predictable and uniform place that a mechanicus could manipulate through a clearly defined set of five mechanical powers (or *potentia mechanicæ*). Again, these mechanical powers were no novel inventions of the Bureau. They are were first discussed in the classical works of Heron of Alexandria, and consisted of *vectis* (the lever), *peritrochium* (the wheel and axle), *trochlea* (the pulley), *planum inclanatum* (the inclined plane) – to which also *cuneus* (the wedge) belonged – and *cochlea* (the screw).[101] In Brandt's, but not Schultze's, manuscript, the last three powers were illustrated by enumerated drawings in the margins. These drawings visualised the mechanical powers using geometric lines, figures and shapes, and some of them were drawn in a simple perspective (see Figure 5–7).

This manuscript shows how the mechanics taught in the Bureau was based on a long classical tradition. However, by adopting this tradition men in the Bureau could claim a specific position in the state. Geometry and perspective drawing were ways for the men of the Bureau to visualise mechanical movement, and mechanics was a way for them to understand human intervention in a geometrically perceived nature. Similarly to how nature was imagined to be balanced and geometrically intelligible, human intervention through mechanical means was easy to identify and to classify. When transcribing the "Short introduction to […] mechanics", an auscultator would come to understand mechanics to be a technique of bringing about ordered change, or a set of *expected transgressions* of a natural balance. However, in the early 1700s this classical geometric and mechanical

---

Brandt's manuscript is paginated, Schultze's is not. In the following notes, the pages thus refer to Brandt's version. A noteworthy difference between the two is that Brandt's manuscript states that the text on mechanics is based on "the writings of learned men as well as individual experience" (in original: "Af lärde mäns skrifter, så wähl some gen förfarenhet samman hämtad", whereas Schultze's does not; Brandt: "Kort handledning till Driftkonsten eller Mechaniken", 246; Schultze: "Kort handledning till Driftkonsten eller Mechaniken".

[100] "*potentias mechanicas*"; Brandt: "Kort handledning till Driftkonsten eller Mechaniken", 247.

[101] Ibid., 263. According to the manuscript, in principle all powers were reducible to *vectis*. For a discussion on the classical mechanical powers, found originally in Heron's *Mechanics*, see Sophie Roux and Walter Roy Laird: "Introduction", in Walter Roy Laird and Sophie Roux (eds): *Mechanics and natural philosophy before the scientific revolution* (Dordrecht, 2007), 4; Mark J. Schiefsky: "Theory and practice in Heron's Mechanics", in Roy Laird and Roux (eds): *Mechanics and natural philosophy*, 16–17.

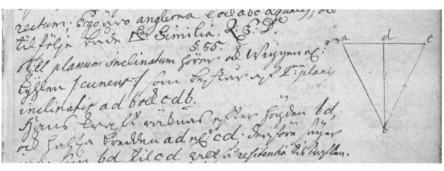

Figure 5, 6 & 7. Illustrations of mechanical powers, from the margins of Brandt's handwritten manuscript (UUB A 28). From top to bottom: the pulley, the inclined plane (in the form of a wedge) and the screw (Photo: Uppsala University Library)

knowledge came to be linked to a specific interpretation of nature, which would make mathematics even more potent as a foundation for a community of men in the Bureau.

"Put away all the prejudices of childhood"

In the manuscript on "*Philosophia naturali*" (i.e., natural philosophy) from Brandt's collection of handwritings, geometric perception and mechanical power converged in a mechanised and geometrised natural philosophy and a philosophy of mind. From studying this manuscript, it is possible to understand the symbolic role of geometry and mechanics in the Bureau of the mid-eighteenth century.

The text started by laying down some requirements put on the would-be natural philosopher (i.e., the young auscultator):

> He who wishes to make positive progress in *philosophia naturali* must first put away all the *prejudices of childhood* [*præjudicia infantiæ*], or the conclusions and opinions of the state of natural things that one has formed in one's youth, because they are brought about by our deceitful senses, which never present a thing such as it is in itself, according to its correct attributes.[102]

The study of natural philosophy was presented as yet another process of maturity, much like how the authors in the previous chapter discussed mathematics, crafts and mechanics. To put away your childish thoughts of the world – based on everyday experience – and to adopt the mathematical worldview shared by the officials was to continue the progress from adolescence into manhood that ideally had begun with exercises of mathematics and crafts during one's childhood (see Chapter 2). To adopt the perceptual framework of the community was also to take on a certain authoritative position shared by the men of the Bureau. The manuscript continued to establish how "we" viewed the world. In a modest tone, this "we" did not denote the officials of the Bureau such as they were, but what they had been before becoming participants in its community of men. "We" were untrained and young, and "our" minds were full of prejudices put there by undisciplined senses:

> We consider the Earth to be larger than the sun and the stars; the sun and the moon to be equally large; the stars to be pretty small, and to be fastened like nails to the sky; that the sky revolves around the earth in 24 hours; and that there is more *matter* in a leaden dice or in that of any other metal, than in a spring of air or of fire.[103]

---

[102] "Den som will giöra lyckligt framsteg uti *Philosophia naturali*, måste först bordtlägga alla *præjudicia infantiæ*, el:r de sluut och omdömen, som man i ungdomen fattadt om de naturlige tingens beskaffenheter, eftersom de äro förordsakade af wåra bedräglige sinnen, som aldrig föreställa oss ett ting, såsom det i sig sielft är till sina rätta egenskaper."; ibid., 334.
[103] "Wij tycka att jorden är större än sool och stiernor, at sool och måne äro lika stora, at stiernorna äro ganska små, och sitta som spikar på himla hwalfwet; At himmelen hwälfwer sig

118

In the manuscript, auscultation was a process of learning through encul-turation that was also a process of maturity. The Bureau's officials were disciplined and rational men, contrasted to imagined uninformed others who based their view of the world on everyday experience.[104] It also implied that all men of the Bureau had gone through this process: from adolescent prejudice to a mature or manly way of viewing nature. This manly view of nature was an eclectic one, which balanced between Cartesianism, classical authorities and the empirical study of nature. The mature man of the Bureau was hence a man who could compare and conciliate such opposed philo-sophical positions. The text pointed out that, although the old philosophers had weeded out many prejudices, it was "the new *philosophy*" that taught the men of the Bureau "not to trust in *sensus*, before it has been examined by *ratio*." The manuscript referred to the principle of Cartesian doubt ("*dubitatio cartesiana*") and the view that the senses should not be trusted too much – "*non nimis sensibus fidendum est*" – as important foundations for philosophical investigation.[105]

In spite of its praises for Cartesian doubt, the manuscript's philosophical method was not Cartesian through and through. Instead it proposed an eclectic philosophy of nature, mind and perception, which was heavily based on Cartesian mechanism but also reconciled a broader set of authorities.[106] As such, the manuscript is very similar to Rydelius' work discussed in Chap-ter 2. In the process of transcribing the manuscript on natural philosophy, the young auscultator would learn that he perceived and made nature intelli-gible through three modes: *purus intellectus*, *sensus* and *imaginatio*. *Purus intellec-tus* occurred when the soul thought in non-visual ways. Thus, through this rational perceptual mode the soul could "perceive the spirits, and their at-tributes and characteristics". Second, the manuscript described how "imagi-nation occurs, when the soul *applies* itself to the brain, and there simulates figures and all sorts of things, which it then observes; as when one imagines a circle, triangle, machine etc." Imagination was thus presented as a visual mode of thinking, a mode of discernment that joined perception and cogni-tion and was brought about by the interaction of mind and body, or soul and brain. The manuscript described how these faculties were exercised and disciplined through geometry and mechanics. Thus, the manuscript pro-

---

omkring jorden på 24 timmar; at det är mehr *materia* uti en tärning bly el:r annan *metall*, än som uti lika stort skråf af luft el:r eld"; ibid., 334–5.

[104] The manuscript thus dismisses the Aristotelian use of everyday experience as a basis for natu-ral knowledge. For a further discussion of Aristotelian and experimentalist views of experience, see Dear: *Discipline & experience*, esp. 11–14, 228, 246.

[105] "nya *Philosophien*"; "ingen Lit på *Sensus*, för än *ratio* den har *examinerat*."; Brandt: "[On Philoso-phia naturali]", 335.

[106] This philosophical method, based on mechanism and a geometrical method, bears many similarities to the "style of thought" of the circle around the *Collegium Curiosorum* discussed by David Dunér. See Dunér: *The natural philosophy of Emanuel Swedenborg*, 44–7; See also Fors: *The limits of matter*, 85.

posed a framework for understanding and structuring perceptions, which also made the auscultators' studies of geometry and mechanics meaningful. Mathematical exercises disciplined young men and developed their *imaginatio* to conform to that expected of the officials in the Bureau. Through this mathematically exercised imagination, the soul and the brain could represent "absent matter" (*"materiale absens"*). Consequently, disciplined imagination made creative but orderly innovation possible. Third, "sensations occur when something touches the senses externally".[107]

From this theory of the mind, the text continued to describe the world that was perceived through these modes of perception. The manuscript stated that nature was infinitely divisible, and filled with natural bodies, understood as "extended in length, width, and depth".[108] Bodies were divided by motion into particles so small that not all could be perceived by *sensum organis*.[109] This world of small particles was in constant motion, and this motion was the source of all "diversity that is found in nature."[110] Thus, the quick motion of the water particles and the relative inactivity of the dirt particles explained the differences between dirt and water.[111] This world, the manuscript suggested, was made intelligible through its similitude to machines:

> We therefore think of this world as great *machine* composed by inimitable art, of innumerable other lesser *machines* and artifice; like a clockwork is governed by its wheels and other properties, for which nothing more is needed [to understand it] than its rest, figure, position, motion and composition.[112]

This *"Philosophia naturali"* could be described as a mechanical philosophy, which explained everything "in terms of the size, shape, and motion of the parts that make it up, just as we explain the behavior of a machine."[113] As pointed out by Daniel Garber, mechanical philosophy, which developed in Europe of the second half of the seventeenth century, should be distinguished from the knowledge of mechanics as such.[114] The mechanics found in Heron, Aristotle, or in the "Short introduction to [...] mechanics" of

---

[107] "betrachtar siälen andarna, och deras *attributa* och egenskaper", "*Imagination* skier då, när siälen *applicerar* sig till hiernan, och *fingerar* der *figurer* och allehanda ting, dem hon då betrachtar; såsom när man *imaginerar* sig en *Cirkel, Triangel, Machin* etc."; Brandt: "[On Philosophia naturali]", 338. "materiale absens"; "*Sensationer* skier när något rörer sinnena utwertes"; ibid., 339.

[108] "*in longum, latum et prosundum extensa*"; Brandt: "[On Philosophia naturali]", 340.

[109] Ibid., 343–5.

[110] "*motus* är ordsaken till alla *diversiteter*, som i *nature* förmärkas."; ibid., 345.

[111] Ibid., 357.

[112] "Wi ansee fördenskull denna werlden såsom en stor *machine*, med oförliknelig konst sammansatt, af oräknelige andre smerre *Machiner* och Konststycken; lika som ett uhrwärk regeras af sina hiuhl och andre tillhörigheter hwartill intet annat behöfwes än desamma wihla, storlek, figur, ställning, rörelse, sammanhang"; ibid., 359.

[113] Daniel Garber: "Remarks on the pre-history of the mechanical philosophy", in Sophie Roux (ed): *The mechanization of natural philosophy* (Dordrecht, 2013), 7.

[114] Ibid., 8–9.

Brandt and Schultze's manuscripts was not a doctrine of nature and were thus not in conflict with Aristotelian natural philosophy. However, the manuscript on "*Philosophia naturali*" explicitly described geometry and mechanics as the basis of a mechanical philosophy of nature, which in turn was presented as the understanding of nature expected of a mature man of the Bureau.[115]

In the manuscript, mechanics and geometry thus merged into an interpretative structure by which the auscultators could imagine the world through the products of their own work. The machines found in the mines governed by the Bureau became mediators between everyday work, nature and the political order of the early modern state. For the officials of the Bureau, mathematics was an exercise that fostered men who could imagine what was not plainly visible, who could see through superficial surfaces to discern truths hidden in nature's depth. What could be more suitable for officials in charge of policing the unearthing of minerals for the benefit of the state? Such mathematical men of metals imagined themselves capable of brining about change and uphold order in the body politic, as well as able to tame nature for the benefit of the *publicum*. By diligently transcribing manuscripts on geometry, mechanics and physics in the Bureau's archive, a young man would be enculturated into this community, and would become a part of its geometric and mechanistic vision.

Conclusions. The mathematics of a balanced state

The Bureau of Mines was a community of diverse forms of knowledge: most importantly of assaying, law and mathematics in a broad sense. This chapter has studied the process of how, over the first half of the eighteenth century, a community formed around geometry and mechanics. By studying the auscultatory system of the Bureau of Mines, I have shown the various ways in which these forms of knowledge were a part of a community of mining cameralists, and how, by the mid 1700s, mathematics gained an important symbolic role for the Bureau's officials. The Bureau was a context in which the imagined intergenerational relationships of mechanics and mathematics, discussed in Chapter 2, were institutionalised into a community of practice. Through a web of relational performances, carried out between older officials and younger auscultators, the Bureau reproduced and guaranteed its own permanence. From 1700 to 1750, the structure of this web changed, as did the role of mathematics and mechanics in the relational performances that constituted it. First, the Bureau's system of auscultation expanded; the base of recruitment became more firmly delimited to

---

[115] This manuscript thus also reinforces Fors' point that Brandt was central in "mechanizing" the chemistry of the Bureau; Fors: *The limits of matter*, 90–7.

universities, and especially to Uppsala University. Second, the role of mathematics changed: from having been strongly linked to mining mechanics, primarily performed in relation to the mechanicus Christopher Polhammar, it gained a symbolic role as a common framework of perception. Mathematics became a way for officials to discern connections between natural phenomena in a disciplined uniform way, as well as a means of maintaining controlled intervention in fragile, and ideally balanced, natural and political orders. Although mathematical textbooks described mathematics as a necessary skill for public office as early as the late seventeenth century, by the 1710s, the Bureau's auscultators Georg Brandt and Anders Gabriel Duhre published works that forcefully presented geometry as a basis for all the Bureau's work. These authors envisioned mathematical knowledge as the basis for not only mining mechanics but also for chemistry and the mining enterprise as a whole.

In application letters from 1700 to 1750, young men conformed to, what they anticipated as, the historically contingent norms of the Bureau. They expressed a desire to work diligently as servants submitted to the public will. As such, the applicants aligned themselves to the ideal of the virtuous and public man of the civil service. Whereas the applicants kept professing their submission to the patria throughout the whole time period, by the 1730s, direct references to servitude under the monarch had all but disappeared.[116] In the bureaus of the early modern Swedish state, where trust and subjugation were intimately linked, the position of a servant subjected to the state (i.e., a civil servant) was paradoxically a means to be identified as a free, trustworthy and mobile actor. The new symbolic role of mathematics, which developed in the Bureau from the early 1720s, can also be discerned in applications to auscultate. By the 1730s, a large minority of applicants discussed their mathematical studies without explicitly seeking a career as a mining mechanic. These applicants presented themselves as potential legitimate participants in a community formed around mathematics. As seen in the auscultator Isaac Johan Uhr's *The Characteristics of an Ironmaster*, in the Bureau of the mid 1700s, mathematics was imagined to be an exercise that formed a young man into someone superior to a commoner working with mining and smelting. This mature man of metals carried many similarities to the mechanicus, discussed in the previous chapter. By the ways in which he integrated mathematics into his work, he distinguished himself as different from, and superior to, other men.

After having been accepted, the auscultators were enculturated into the mathematical community of the Bureau: they aligned their minds to its communal way of perceiving and imagining nature and social relationships.

---

[116] In the next chapter, I will study this development in more detail, by narrowing in on the role of royal patronage in the work of a handful of mathematical and mechanical practitioners of the first decades of the eighteenth century.

This process of enculturation involved a number of exercises: listening in to the board, writing protocols, participating in the Bureau's work and travelling both domestically and abroad. The practice of transcribing texts was another way in which the auscultators could align their minds to the norms and practices of the Bureau. These texts included travel reports, mining ordinances and protocols, as well as educational hand-written manuscripts on a variety of subjects. In order to understand the symbolic role that mathematics gained in the Bureau by the 1730s, a handful of copied manuscripts concerning geometry and mechanics are especially illuminating. By copying these texts, an auscultator could internalise the geometric framework of perception. He would learn to see nature as geometric points, lines, shapes and figures; he would also understand how to intervene in a balanced nature through a definite number of mechanical powers; and he would come to see nature as both constantly in motion and as fundamentally balanced. These manuscripts presented the world using metaphors that involved the very machines found in the mines that the Bureau policed, and skills that were part of the Bureau's everyday work. Thus, mechanics became a means by which the community could come to terms with the world through the material artifacts in the mines that they policed. When copying these manuscripts, the auscultators would be enculturated into a community of men who saw themselves as carrying out continuous expected transgressions. Their work consisted of bringing about changes in the Swedish state, in order to perfect its theocratic order. The geometry and mechanics of the Bureau were thus mathematics of a balanced state.

In this chapter, I have discussed performances of mechanics and mathematics that can be considered a system of relational performances, through which men shaped themselves into a coherent hierarchical community. By the mid-eighteenth century, relational performances of geometry and mechanics became an important basis for this community. Through these techniques, the men of the Bureau imagined that they perfected themselves into ideal servants of the *publicum* (i.e., to the king and later to the estates of the parliament). But mechanical practitioners did not only submit to the political power through the state administration. In the ensuing chapters, I turn to two cases connecting mechanics to early modern political order. The first case mainly unfolds during the period of Swedish absolutism of the late seventeenth and early eighteenth centuries; the other is mainly from the constitutional monarchy post-1720. In both these cases, my focus is on the direct relationship between mechanical practitioners and the ultimate political power. The focus of these chapters is the relationship between the mechanicus and the shifting shapes of the *publicum*.

4. Mechanical correspondence. Relational performances of the monarch and the mechanicus

In 1722, the mechanicus Christopher Polhem wrote several letters to Erik Benzelius, the librarian at Uppsala University. In these letters, he commented on the political developments of his time: how the absolute monarch Karl XII had died in 1718 and how the parliament, consisting of representatives of the four estates of the realm (nobility, clergy, burghers and peasants) *de facto* had seized power in Sweden and put an end to the royal autocracy. Polhem mourned how all love for himself, as well as "the love for mechanical science, has fallen asleep together with the blessed king."[1] He explained:

> that I have not [made anything useful for the realm] after the death of the blessed king, is unsurprising. Simply consider what happens in a private household where the master and the children are put under the guardianship of the servants. In opposite to how other nations put the mere theory of mathematical, mechanical and physical sciences in high esteem, here one now scorns both theory and practice.[2]

In Polhem's opinion, the new constitutional regime had disrupted the traditional patriarchal theocracy of which mechanics and mathematics were parts. In another letter to Benzelius, written a month later, he returned to the subject. Reminiscing that they had corresponded for 12 years, he remarked that neither of them would have been able to predict the changes in the realm over the past decade: "at that time the realm existed for the sake of the king, not the king for the realm. [...] Then the *œconomy* and culture of the realm lay severely ill, now basically at death's door".[3] In a third letter the same autumn, he yet again returned to the subject of government:

---

[1] "för Mechanska wettskaper wore tillijka med Salig Konungen afsomnad."; Christopher Polhem: "Letter to Erik Benzelius, 1722-08-09", in Axel Johan Carl Vilhelm Liljencrantz (ed): *Christopher Polhems brev* (Uppsala, 1941), 144.
[2] "att iag intet komit dertill effter Salig Konungenss död åhr intet så aldeless till förundra, när man allenast betrachtar huru det går till i ett privat huss huarest man sätter huss fadren och Barnen under tienst folkets förmy[n]derskap. I stellet andra Nationer gör stort æstime af blotta theorin af Mathematiska, Mechanska och Physicalsk[a] wettskaper, så gör man nu här ett föracht både af theori och practic,"; ibid., 145.
[3] "wijd den tijden war Rijket för kungen skull, nu kungen för Rijket skull. [...] Då låg Riksenss æconomie [sic] och cultur illa siuk, nu reent för döden"; Christopher Polhem: "Letter to Erik Benzelius, 1722-09-02", in Liljencrantz (ed): *Christopher Polhems brev*, 146.

It heartens me to see that the Doctor trusts my ability to serve the fatherland using the simple knowledge I have acquired. But nonetheless, I deeply regret that those who have this trust are few and far between. And because our government these days is such, that nothing is considered good that is not supported by the majority, my work will probably be valued even less.[4]

As part of a heroic narrative of invention, biographers of Christopher Polhem from the nineteenth and twentieth centuries have generally described him as a man ahead of his time.[5] Historians who have discussed his ties with the monarch have generally done so *ex post facto* – that is, his contacts with Karl XII have been interpreted as a sign of what was to come, rather than as formed by a contemporary regime of mechanical work.[6] In relation to such a historiography, the statements in Polhem's letters from 1722 seem out of place. If he comes across in these letters as a man out of his time, it is as someone belonging to the past and not to the future. He had apparently not adapted to the new political conditions of the 1720s; he was a man defined by a regime that was no more.

Polhem was 50 years old when Karl XII died. It is thus unsurprising that he was shaped by the absolutist regime in which he had lived the greater part of his life. Still, few biographers have described him as a man marked by Caroline absolutism.[7] This is a striking fact because few Swedish mechanical practitioners of the eighteenth century were more successful in fashioning themselves in relation to the absolute monarchy. Even his surname, Polhem, was a product of this relationship with the king, who ennobled the mechanicus, formerly known as Polhammar, in December 1716.[8] To make clear that Polhammar/Polhem was transformed by the relationships that I am studying, I will use the name Polhammar up until the point that he was ennobled, and Polhem afterwards.

Although a number of studies have covered the role of eighteenth-century royal patronage of mathematics and natural philosophy, few have focused on the aspect of reciprocal self-fashioning in these relationships.[9] It

---

[4] "Det fägnar mig myket att see H. H. Doct: goda tanka om min duglighet att kuna göra fädernesslandet tienst af de ringa wettskaper iag inhemptad. men beklagar der jempte att de som hafva detta tycket lära wara myket tunt sådda. Och effter wårt regiments wässende nu för tijden åhr sådan, att ingen ting hålless för godt som icke har den större delenss bijfall, så lärer mina förrättningar så myket mindre blifva ansedda,"; Christopher Polhem: "Letter to Erik Benzelius, 1722-11-05", in Liljencrantz (ed): *Christopher Polhems brev*, 160.

[5] For examples of such narratives, from the 1900s as well as from today, see page 41, note 5.

[6] For example, Bring describes the relationship between Polhem and the king as founded on their mutual visions for the future; Bring: "Bidrag till Christopher Polhems lefnadsteckning", 57.

[7] Lindqvist has pointed out the importance of royal ties for Polhem's career in the Bureau of Mines, in Lindqvist: *Technology on trial*, 99.

[8] Bring: "Bidrag till Christopher Polhems lefnadsteckning", 49.

[9] For an in-depth study of patronage of mathematics and natural philosophy in absolutist regimes, see Biagioli: *Galileo, courtier*. S. N. Eisenstadt and Louis Roniger has discussed patron–client relations as "models of structuring the flow of resources and of interpersonal interaction and exchanges in society"; "Patron–client relations as a model of structuring social exchange", *Com-

is this aspect of patronage that concerns me here. This chapter focuses on the letter correspondence between Christopher Polhammar (1661–1751), Emanuel Swedenborg (1688–1772), Karl XII (1682–1718) and the king's secretary, Casten Feif (1662–1739). These four men were part of a complex web of patronage; over the first two decades of the 1700s, they established heterogeneous positions of authority by chiseling out highly interdependent identities.

Early modern monarchs were staged as symbols of a specific social order, in ceremonies, portraits and architecture. I argue that mechanics was another means of performing such royal symbolism. Mechanistic conceptions of the political order formed a structural framework in which the mechanicus and the monarch corresponded. In their letter correspondence, Polhammar and the king could act tactically in relation to this framework, and reciprocally mould each other into what was expected of a monarch or mechanicus.

The monarch perceived

Then who, or what, was the absolute monarch? First: absolutism was not as absolute as it has been made out to be. Early modern "absolute" monarchs ruled through collaboration with, and by balancing, socially powerful elites both at court and in the provinces.[10] Similarly, over the past decades, several scholars – especially on the French absolute monarchy of Louis XIV – have examined the rites and social practices of absolutism. They have approached the absolute monarch as a symbol that was fabricated through representations in various media.[11] From 1680, when Karl XI introduced absolutism in Sweden, similar symbolism of an imagined strong and centralised monarchy, modelled after a French example, also became common in Sweden.[12]

---

parative studies in society and history 22:1 (1980), 56. On the role of patron–client relationships in eighteenth-century Sweden, see Patrik Winton: *Frihetstidens politiska praktik. Nätverk och offentlighet 1746–1766* (Uppsala, 2006), 29–30. On patron–client relationships in eighteenth-century Swedish chemistry, see Fors: *Mutual favours*, 200–1. A common interest for these studies is patron–client relationships as a means of reciprocal favours from which the parties benefitted. However, they focus less on how the parties were fashioned through such hierarchical relationships. Karin Sennefelt has a perspective more similar to mine, when discussing how eighteenth-century men in Stockholm were transformed into equals or patron–clients through the transformative practice of alcohol drinking. She highlights how these men transformed each other through encounters where social hierarchies were pivotal; Sennefelt: *Politikens hjärta*, 150.

[10] For an overview of the research that has shown the collaborative aspects of absolutism, see William Beik: "The absolutism of Louis XIV as social collaboration", *Past & Present* 188:1 (2005), 195–7.

[11] Jean-Marie Apostolidès: *Le roi-machine. Spectacle et politique au temps de Louis XIV* (Paris, 1981); Peter Burke: *The fabrication of Louis XIV* (New Haven, 1992), 151–178.

[12] The self-fashioning of early modern Swedish monarchs has until recently been a neglected field. Older studies include Ragnar Josephson: "Karl XI och Karl XII som esteter", *Karolinska förbundets årsbok* (1947), 7–67; Allan Ellenius: *Karolinska bildidéer* (Uppsala, 1966); Kurt Johannes-

The symbol of the king was, moreover, a symbol of the realm. From medieval times and into early modernity, English political theology embraced a theory of the king's two bodies – a personal body of a specific man and a symbolic body that could not die: a body politic. Being made up of two bodies, the sovereign existed in between the individual and the collective. The parts of the monarch's political body were the subjects of the realm, and the monarch thus embodied the relationships between sovereign and subjects in a hierarchical society. In other words, early modern monarchs were not just individuals in a line of succession. They also represented a continuous office, a symbol or a persona that corresponded to a specific social order. In a number of early modern royal ceremonies, this relationship between the king's two bodies was presented to the subjects of the realm. Ernst Hartwig Kantorowicz has identified the funeral ceremony as one such forum for staging this relationship, whereby one royal individual succeeded another, but the office remained constant.[13] Moreover, Abby E. Zanger has shown how royal bodies were presented in French royal marriage ceremonies. While both marriage and funeral ceremonies promoted the dynasty, the former did not promote the "absolute oneness" of the royal bodies. Instead, marriage "is always necessarily dialogical, even if there is a power struggle between single and dual perspectives within its symbolic field." Hence, the relationships sealed in royal ceremonies did not only guarantee royal continuity in spite of mortality, but furthermore joined together members of royal families.[14] Similarly, research on early modern courts has shown how monarchs were not singular: the courts were integral to the relationships between the monarchs and their European subjects. In a court, the king's personal power – manifested through favours and patronage – emanated through chains of subjects at different degrees of closeness to the king.[15]

---

son: *I polstjärnans tecken. Studier i svensk barock* (Stockholm, 1968). Recently, a growing number of studies have appeared that examine portraits, ceremonies and architecture as performances and staging of monarchs. See for example: Snickare: *Enväldets riter*; Martin Olin: *Det karolinska porträttet. Ideologi, ikonografi, identitet* (Stockholm, 2000), 49–51; Snickare: "Shaping the ritual space"; Kekke Stadin: "The masculine image of a great power. Representations of Swedish imperial power c. 1630–1690", *Scandinavian Journal of History* 30:1 (2005), 61–82; For a discussion of, and comparison between, the coronation ceremonies during the Swedish absolutism and constitutional monarchy, see Nordin: *Frihetstidens monarki*, 59–83.

[13] Ernst Hartwig Kantorowicz: *The king's two bodies. A study in mediaeval political theology* (Princeton NJ, 1957), 7–23.

[14] Abby E. Zanger: *Scenes from the marriage of Louis XIV. Nuptial fictions and the making of absolutist power* (Stanford CA, 1997), 8.

[15] For an overview of the research of the early modern European court, see Persson: *Servants of fortune*, 3–8.

Presenting the royal gaze

The monarch was staged through royal performances made for heterogeneous audiences throughout the realm and abroad. It was a symbol, which was not only perceived: an imagined power of perception was at the core of subjects' understanding of the king. In principle, the king, as the head of the body politic, saw and understood every event in the social body. The monarch's gaze was thus a guarantee of order and correspondence between superiors and subordinates.

Wakefield has discussed a letter from an otherwise unknown man named Gottlob Christian Happe to Friedrich II of Sachsen-Gotha, dated 7 April 1717. In his letter, Happe described to the sovereign an architectural invention for fabricating a good social order. Happe imagined chambers in which the council of the realm would meet. A network of tunnels would connect the sovereign's quarters to cabinets, or "loges", adjacent to these chambers. From these loges, it would be possible for the sovereign to observe the work of his subjects, without their knowing whether he was there or not.[16] Wakefield points out that there is "no evidence [...], that Happe's loge was ever built or that any prince even bothered to read his book." It has "a whiff of desperation about it," the product of an impoverished scholar seeking "patronage and preferment [by constructing] elaborate visions of control and order."[17]

Still, Happe's invention is helpful in highlighting how subjects imagined sovereigns in eighteenth-century northern Europe. His work can be related to the tradition of renaissance "machine books", filled with pictorial and literary descriptions of machines, through which mechanical practitioners presented their work to sovereigns and other patrons. As pointed out by Jonathan Sawday, such works did not always present workable machines.[18] Often the designs were fanciful, made to be read rather than used as designs for construction. Machine books came to inhabit "fantasy as much as reality." Early modern mechanical practitioners were not only "responding to modern design criteria." Their designs were also a "glimpse [...] of a more idealized world" presented to a princely audience.[19] Happe's design was one such performance, which allowed the monarch to imagine himself as an all-knowing perceiver. As such, it is an example of how inventions, imagined or real, could be a means of shaping and intertwining mechanical practitioners and sovereigns of early modern absolutist states.

The sovereigns of seventeenth- and eighteenth-century Europe were increasingly dependent on developing state bureaucracies. If the monarch were to be regarded a symbol of a body politic, the cameral administration

---

[16] Wakefield: *The disordered police state*, 14.
[17] Ibid., 15.
[18] Sawday: *Engines of the imagination*, 83–6.
[19] Ibid., 97.

was identified as its hands and eyes. In the early eighteenth century, cameralists, and cameral knowledge, were presented as ways of attaining fiscal means for the monarch and of developing the economy of the realm. As discussed in the preceding chapters, mathematics and mechanics were inherent parts of such cameralism, which provided a means of governing early modern states. Jay Smith has shown how nobles in the bureaucracy of absolutist France, when discussing their merits in written text, "made frequent reference to the person of the king". It was, Smith argues, in the light of the "sovereign's gaze" that "the nobility's unique qualities would inevitably be revealed and rewarded."[20] But this gaze required a government that consisted of more than a king. A stable absolute monarchy needed a bureaucracy that could outlive an individual king, while serving as a manifestation of the monarch in his official capacity.[21] Consequently, Louis XIV wished to reaffirm his absolute power, and to circumscribe the power of the nobility, by making the traditional "personal modality of service" systematic. That is, he wanted the bureaucracy to act according to certain principles of merit, rather than through personal recommendations. Ironically, the ambitions of Louis XIV to make his position absolute also laid the groundwork for a culture of impersonal service increasingly autonomous of the king.[22]

Mechanics of absolutism

Common to these studies – both of ceremonial performances and of the monarch's gaze – is the understanding that the sovereign was fashioned in relation to his subjects. As discussed in the introduction, by the late seventeenth century, mechanics could make such relationships intelligible when it was integrated with other means of conceptualising social relationships. The relations that constituted the social order were often conceptualised through mixed metaphors of machines and organic bodies. According to this line of thinking, society was a body, which was a machine. The machine in turn thus became a mediator between conceptions of bodies and society.

In Continental Europe and Scandinavia, mechanics and geometry were linked to an authoritarian political order. The English mechanical philosopher Thomas Hobbes saw parallels between the force behind geometrical inferences and the order ensured by an absolute power, a Leviathan.[23] Similarly, German cameral writers of the eighteenth century legitimated the

---

[20] Smith: *The culture of merit*, 4–5; See also Alder: *Engineering the revolution*, 49.

[21] For a discussion of the relationship between the monarch and the state apparatus of early modern Sweden, see Lindberg: *Den antika skevheten*, 12.

[22] Smith: *The culture of merit*, 264.

[23] Marshall Missner: "Skepticism and Hobbes's political philosophy", *Journal of the history of ideas* 44:3 (1983), 410–11; Shapin and Schaffer: *Leviathan and the air-pump*, 101–7, 153. Likewise, Sawday points out that for "Hobbes, life, whether it was the life of individuals or societies was essentially 'motion', a mechanical phenomenon, which could be investigated with the clockmaker's precision and skill"; Sawday: *Engines of the imagination*, 246.

power of the sovereign through machines and mechanics. In 1736, 4 years before he claimed the throne of Prussia, King Friedrich II (1712–86) argued that a politician should act "as an able mechanic". Like a mechanicus – who should not only be "content with examining the exterior of a clock" but of whom it was expected "that he opens it, that he examines the springs and wheels in it" – an able statesman should "know the permanent principles of the courts, the springs of the politics of each prince, the sources of all events."[24] Later in the eighteenth century, the German cameralist Johann Heinrich Gottlob von Justi (1717–71) described the sovereign as "the governor of the machine of the body politic [*Staatskörpers*]."[25] In the geometric and mechanical vision of authoritarian order, which merged organic and mechanistic metaphors, the king was the head of the social clockwork, as well as the mechanicus who guaranteed the movement of its parts. The products of mechanical work, and the dominion of mechanici over craftsmen and nature alike, became a means of arguing for a centralist vision of society, the rule of God and the absolute monarch.[26] In return, this conception of political order legitimated certain approaches to the construction of mechanical machinery, where ideal machines were perceived as autonomous and hierarchical. While mechanics made an authoritarian order intelligible and legitimate, this political order in return made mechanics meaningful.[27]

A clear manifestation of this reciprocal yet asymmetrical bond was a representation of society in the form of an automaton, presented to Frederik IV of Denmark–Norway around 1700 (see Figure 8). The automaton was a portable machine model that consisted of two parts: the left one representing the machines and men of mining, and the right one representing the governmental hierarchy of the same mining production. King Frederik could make both sides of this automaton-society move by turning appropriate knobs on the back of the machine. According to the vision embodied in this piece of machinery, society was a mechanical body, and its subjects were wheels and springs. The mechanicus both constructed the model of a well-ordered society (being the gift giver) and was represented in the machine by his constructions (on the top left-hand side). Similarly, the

---

[24] "Comme un habile mécanicien"; "ne se contenteroit pas de voir l'extérieur d'une montre"; "qu il l'ouvriroit, qu'il en examineroit les ressorts et les mobiles"; "un habile politique s'applique à connoître les principes permanens des cours, les ressorts de la politique de chaque prince, les sources des événements"; Friedrich II of Prussia: "Considérations sur l'état présent du corps politique de l'Europe", *Oeuvres posthumes de Frédric II, roi de Prusse* vol. 6 (Berlin, 1788 [1736]), 4.

[25] "der Regierer von der Maschine des Staatskörpers."; Johann Heinrich Gottlob von Justi: *Der Grundriss einer guten Regierung in fünf Buchern* (Frankfurt and Leipzig, 1759), 329. On Justi's machine state, see Wakefield: *The disordered police state*, 34.

[26] Raeff: "The well-ordered police state", 1229–30; Mayr: *Authority, liberty & automatic machinery in early modern Europe*, 115–21.

[27] Schaffer: "Enlightened Automata", 136; Mayr: *Authority, liberty & automatic machinery in early modern Europe*, 102–3, 115–21.

Figure 8. The cameralist order of society represented in the form of a portable machine model, offered to King Frederik IV of Denmark and Norway around 1700. (Photo: Tor Aas Haug/Norsk Bergverksmuseum)

monarch was both a part of the machine (residing at the top right-hand side) and guaranteed the order of the whole device by turning the knobs in the background (as the owner of the machine itself). In this second sense, the monarch inspected the machine with penetrating vision and controlled the circulation of its parts by turning the knobs.[28]

Perhaps an even more concrete example of the role of mechanics in an absolutist order can be found in the correspondence of a more or less unknown contemporary of Christopher Polhammar: Birger Elfwing (1679–1747). In a plea for monetary help, on Christmas Eve 1716 Elfwing composed a letter to Karl XII, in which he requested support for his various mechanical projects:

> I do not believe it possible to complete [these and other inventions] using private means. Instead, I most subserviently submit to the most understandable grace of Your Royal Majesty, who has cared for the progress of several other sciences through subsidies and promotions.[29]

In his letter, Elfwing wished to present his "most subservient zeal for his Royal Majesty and the military service of our fatherland". He could not afford to fund his work himself, and therefore asked for the Majesty's help. Simultaneously, he presented himself to the king as someone with an interest in useful arts and knowledge. Furthermore, he professed an emotional connection to the king. He wished that the "utility of [these inventions], with the help of God, would soon be of assistance to Your Royal Majesty". He "considered it to be of my greatest happiness" to provide "that which is of agreeable use to my most gracious king."[30]

Elfwing attached to his letter an engraving that portrays a mechanicus showing an invention to a would-be patron (see Figure 9). The picture displays both a specific agricultural machine, a threshing-mill run by wind power, and the social relationships of which Elfwing imagined his mechanical work to be a part. The engraving shows the inventor, spelled out as "B. Elfwing: inv", standing next to his patron. The inventor points towards the machine, while facing the patron for whose benefit he has made his invention. In the background, two women are working in the rural setting where

---

[28] For a more detailed description of this machine model, see Christensen: *Det moderne projekt*, 74. Peter Dear has argued for the similarities between seventeenth-century kingship and scientific authority, in ways that underline my point, in Dear: "Mysteries of state, mysteries of nature", 214–20.

[29] "ser jag mig af egen förmögenhet sådant eij hinna utföra, utan allerunderdånigst hemställer Eders Kongl. May:ts högstbegripliga nåd, som om åthskillige andra Scientiers progress med understöd och befordran dragit försorg"; printed in Ernst E. Areen: "Birger Elfving. Hedemora gevärsfaktori och Furudals styckebruk", *Svenska vapenhistoriska sällskapets årsskrift* 3 (1931), 70–1.

[30] "min underdånigste Zele för Eders Kongl. May:ts och Fäderneslandsens Krigs Tienst"; Areen: "Birger Elfving. Hedemora gevärsfaktori och Furudals styckebruk", 68. "att nyttan der af, med Guds hielp, skall snart komma Eders Kongl. May:t till tienst"; "skattandes det för min högsta lycka"; "sådant, som wore min Allernådigaste Konung till behagelig nytta."; ibid., 71.

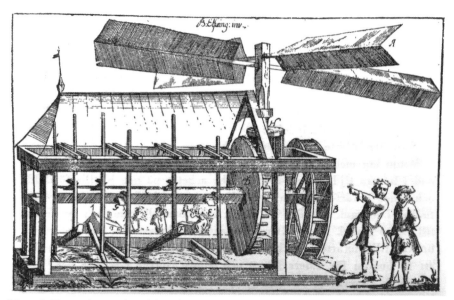

Figure 9. Engraving enclosed in a letter from Birger Elfwing to Karl XII. (Reproduced from Ernst E. Areen: "Birger Elfving", 19)[31]

the machine is supposed to be running. The engraving can be interpreted as an illustration of mechanics' pretentions to relevance in wide parts of society. It portrays an order in which machinery, mechanical knowledge and political power coexisted, while depicting how relationships of superiority and subordination mattered to early modern mechanics.

The positioning of the actors by depth aligns with the power relationships of the absolutist state: the king in the foreground, the mechanical practitioner right behind him, then the machine, and, finally, far in the background and framed by the machine as if part of it, the peasant women. In the horizontal plane, the mechanical practitioner is positioned as a trusted mediator between the king, on the one hand, and a hierarchical order of production of workers and wooden cogwheels on the other. In turn, the machine is placed as a mediator between the women, working in the background, and the two men to the right. Elfwing's letter and engraving hinted at royal patronage as something that involved more than the granting of economic aid to useful clients. Patronage also had a symbolical and identificatory function: it was another forum where royal power could be staged. By enacting such relationships, monarchs and mechanici were intertwined and shaped each other in the eyes of their contemporaries.

Thus, in early modern Europe, patron–client relationships between monarchs and mechanici were not only ways for clients to attain resources

---

[31] Areen gives a reference to a letter in the collection "Diverse militära inventioner" in the Military Archives of Sweden (Krigsarkivet); Ernst E. Areen: "Birger Elfving", 68. However, this collection does not exist anymore in that archive, and the letter has not been found. Therefore, I reproduce the illustration from Areen's article.

for mechanical work. As machines and bodies mediated between orders of production, nature and society, relationships between monarchs and mechanical practitioners were given a range of connotations. For those who conceived society as a machine-body, the machine made the monarch's absolute power and gaze intelligible. The social epistemology of early modern mechanics formed a structural framework that linked the work of the mechanicus to the monarch's exercise of power. Therefore, through the staging of mechanical interest and patronage, the monarch could present himself as a powerful and perceptive sovereign to subjects who shared these mechanistic views. As the meaning of mechanics and royal power corresponded structurally, letter correspondence between kings and mechanical practitioners provided tactics for both parties to fashion themselves in relation to what was expected of a monarch or a mechanicus.

The mechanici, the monarch and the secretary of state

Karl XII claimed the throne at the age of 15 and ruled the Swedish Empire around the Baltic Sea from 1697 until his death in 1718. Following his father Karl XI, and inspired by Louis XIV, he fashioned himself in the image of an absolute monarch with God-given authority.[32] From the onset of the eighteenth century, Sweden was engaged in the Great Northern War (1700–21), in which Russia under Peter I challenged the Swedish supremacy around the Baltic Sea. In the early 1710s, Karl was in his late twenties, and had waged wars, in several European regions, for over a third of his life. The war effort had strained the resources of the Swedish fiscal military state to its limits, and after some initial success in the field the fortune of war had now definitively turned against the Swedes.

Biographers have generally given little attention to Karl XII's participation in the everyday administration of the realm.[33] Living a great part of the 1700s and 1710s abroad, leading his armies, the king was at times only in sporadic contact with the state apparatus at home. Still, along with the image of Karl XII as a warrior king, biographers from the mid-eighteenth century onwards discussed his apparent interest in, and patronage of, mathe-

---

[32] Gunnar Carlqvist: "Karl XII:s ungdom och första regeringsår", in Samuel E. Bring (ed): *Karl XII. Till 200-årsdagen av hans död* (Stockholm, 1918), 75; Ragnhild M. Hatton: *Charles XII of Sweden* (London, 1968), 79–81; Snickare: *Enväldets riter*, 110.

[33] Åsa Karlsson: *Den jämlike undersåten. Karl XII:s förmögenhetsbeskattning 1713* (Uppsala, 1994), 32. As pointed out by Karlson, the two most prominent biographers only briefly discuss the king's participation in the internal affairs of the realm; Hatton: *Charles XII of Sweden*, 337–49; Sven Grauers: "Karl XII:s personlighet. Försök till analys", *Karolinska förbundets årsbok* (1969), 24–25. Furthermore, both biographers refer to Feif when arguing that the king took an active interest in economic reform.

matics.[34] In the second edition of his biography from 1749, François Marie Arouet de Voltaire commented on the fact that "some persons have wanted to present this prince as a good mathematician". Voltaire was not convinced of these accounts: he considered that "the evidence given for [the king's] knowledge in mathematics is not very conclusive".[35] In the same vein, and in opposition to older celebratory histories of Karl XII as a philosopher monarch, more recent scholars have wished for a more nuanced interpretation of the sources celebrating the king's mathematical skill. In this vein, David Dunér argues that there can be a difference between the private and the public views of a scientist. According to Dunér, letters between mathematicians and monarchs were part of a political game distinct from "private opinions" of scientists and actual scientific practice.[36]

Although such interpretations facilitate narratives of royal patronage that move beyond the literal celebratory level of older historiography and of the historical sources, they are nonetheless highly problematic. Attempts to find sources that provide actors' true private opinions, isolated from a political order, are futile and only generate guesses of essentialist qualities. Then how do we write a history of royal patronage of mechanics, which is critical of celebratory biographies, but which at the same time recognises that we cannot disentangle early modern mathematics from the political orders of which they were a part? In this chapter, I analyse presentations of the king and of mechanici as reflections of neither essential qualities nor a superficial façade. Instead, I see them as parts of integrated performances through which all parties were made and remade. Consequently, I care little about whether or not Karl XII was a competent mathematicus, or whether his clients were sincere in their presentations of each other or the king. Instead,

---

[34] Already in the official biography sanctioned by the Swedish state and written by Jöran Andersson Nordberg in 1740, the king was presented as a competent mathematician; Jöran Andersson Nordberg: *Konung Carl den XII:tes historia* (Stockholm, 1740), 599–601. By the late nineteenth century, Karl Siljestrand described the king as in possession of great philosophical insights; Karl K:son Siljestrand: *Karl XII såsom filosof* (Linköping, 1891), 1. Among twentieth-century historians, Hatton mentions Polhem's and Swedenborg's favourable opinions of the king's mathematical skill, without discussing these remarks further, Hatton: *Charles XII of Sweden*, 430. These descriptions of the king as a mathematician can still be found in more popular biographical literature. For example Bengt Liljegren has described Karl XII as "the mathematician on the throne" (in original: "matematikern på tronen"); Bengt Liljegren: *Karl XII i Lund. När Sverige styrdes från Skåne* (Lund, 1999), 65. In a somewhat later biography he slightly adjusts this characterisation, stating that the king was "interested in numbers" (in original: "intresserad av siffror") but that "he was not a mathematician" (in original: "någon matematiker var han inte"); Bengt Liljegren: *Karl XII. En biografi* (Lund, 2000), 317.
[35] "Quelques personnes ont voulu faire passer ce Prince pour un bon Mathématicien"; "la preuve que l'on donne de ses connaissances en Mathématique n'est pas bien concluante"; François Marie Arouet de Voltaire: *Histoire de Charles XII, roi de Suède, divisée en huit livres, avec l'histoire de l'empire de Russie sous Pierre-le-Grand, en deux parties divisées par chapitres* (Geneva, 1768), 382.
[36] "privata åsikt"; David Dunér: "Sextiofyra och åtta istället för tio. Karl XII, Swedenborg och konsten att räkna", *Scandia* 67 (2001), 231; For a similar analysis, see Hans Helander: "Introduction", in Emanuel Swedenborg: *Festivus applausus in Caroli XII* (Uppsala, 1985), 9–50.

these presentations should be studied as parts of interlaced processes of becoming, through which all involved fashioned themselves in relation to the political order of an absolutist regime.

Such entangled performances of mechanics and royal power can be seen in a substantial correspondence between Karl XII, his secretary Feif, Polhammar and a young Swedenborg.[37] From the early 1710s, these four men were shaped by a chain of relational performances in letter form. They inhabited radically different positions in early eighteenth-century society. Apart from them all being men who were part of the theocratic absolutist order of the early modern Swedish state, they had little in common. Their relationships were based on strict hierarchies of age, social position and wealth. Even the relationships as such did not necessarily have equal importance to all parties. Whereas Polhammar was fundamentally transformed through his relationship with the king, the same cannot be said, to the same degree, about Karl XII. Still, no party was left unaffected by these relationships. They were both means of reciprocal favours and central in processes of self-formation, by which all involved parties were shaped in relation to, and grew dependent on, each other.

Polhammar – Swedenborg. The groundwork of an intergenerational relationship

The starting point of the correspondence studied here was the *Collegium Curiosorum*. The *Collegium* was a short-lived scientific society in Uppsala, founded in 1710 by a group of scholars who held meetings in 1710–11, while the university was closed because of a plague outbreak.[38] Its initiator, Benzelius, was part of a European "republic of letters" and corresponded with scholars such as Hans Sloane and Leibniz.[39] Benzelius and the *Collegium* also corresponded with Polhammar on matters of mechanics, mathematics and natural philosophy, and discussed Polhammar's letters at its meetings.[40] Emanuel Swedenborg was the younger brother-in-law of Benzelius.

---

[37] Emanuel Swedberg changed his surname to Swedenborg in 1718, when he was ennobled as an honorary sign for his father's work as a bishop. For the sake of consistency, I use his latter name continuously throughout the chapter. This might appear inconsistent when compared to the way I name Polhammar, but whereas there is an analytic point of discussing him as "Polhammar"—as the name Polhem was a result of the processes that I study—Swedenborg's name change is not a result of the processes studied here.

[38] The *Collegium* is discussed in detail in Dunér: *The natural philosophy of Emanuel Swedenborg*, 41–7 Hildebrand: *Kungl. Svenska Vetenskapsakademien* vol. 1, 81–135; see also Lindqvist: *Technology on trial*, 119–21

[39] For a general discussion of the scholarly republic of letters, see Dena Goodman: *The republic of letters. A cultural history of the French Enlightenment* (Ithaca, 1994), 1–11. On Swedish scholars' relation to the republic, see Fors: *Mutual favours*, 8–9.

[40] On Christopher Polhem and the *Collegium Curiosorum*, see Axel Johan Carl Vilhelm Liljecrantz: "Polhem och grundandet av Sveriges första naturvetenskapliga samfund. Jämte andra anteckningar rörande Collegium Curiosorum. I", *Lychnos* (1939), 289–308; Axel Johan Carl Vilhelm

11 years old, Swedenborg matriculated at Uppsala University and he stayed there for 10 years. After graduating, in the summer of 1709, he started planning a European tour to London, Paris and Leiden. At the same time, he started to discuss mathematics and Polhammar with his brother-in-law.

On 13 July 1709, Swedenborg sent a letter to Benzelius about his tour, asking him for letters of recommendation to the librarian's "acquaintances in England". Swedenborg "thought it advisable to choose a subject early, which I might elaborate in course of time, and into which I might introduce much of what I should notice and read in foreign countries." The subject he wished to pursue was mathematics, which he desired to "turn to some practical use, and also to perfect myself more in".[41] While awaiting his departure, Swedenborg compiled a work *of things discovered and to be discovered in mathematics*, or, what is nearly the same thing, *the progress made in mathematics during the last one or two centuries.*" This work would include "all branches of mathematics". To this compilation, Swedenborg wished to add the inventions of Polhammar, as an ornament of the whole work.[42]

In March the following year, Swedenborg wrote to Benzelius again, stating that his greatest desire was "to obtain some information respecting the plan now being discussed here, of my staying with Polhammar."[43] Swedenborg used his influential family network in order to be introduced to Polhammar and invited to stay with him on his estate of Stiernsund; after some bartering, Polhammar accepted Swedenborg's proposal.[44] Polhammar's decision was not only a result of Swedenborg's influential contacts: he was also convinced by Swedenborg's presentations of mechanical skill. In a letter to Benzelius, Polhammar professed that he was:

> extremely well pleased that he [Swedenborg] came here, […] and as we were pleased and satisfied with one another, his desire could be gratified without any difficulty; especially when I found him able to assist me in the mechanical undertaking which I have in hand, and in making the necessary experiments.[45]

Polhammar portrayed Swedenborg as "a quick and intelligent person".[46] Similarly, Swedenborg described Polhammar as "so great a man, one such as

---

Liljecrantz: "Polhem och grundandet av Sveriges första naturvetenskapliga samfund. Jämte andra anteckningar rörande Collegium Curiosorum. II", *Lychnos* (1940), 21–54.

[41] Emanuel Swedenborg: "Letter to Ericus Benzelius, 13 July 1709", in Rudolf L. Tafel (ed): *Documents concerning the life and character of Emanuel Swedenborg* vol. 1 (London, 1875), 200.

[42] Ibid., 201.

[43] Emanuel Swedenborg: "Letter to Ericus Benzelius, 6 March 1710", in Tafel (ed): *Documents* vol. 1, 202.

[44] Dunér: *Världsmaskinen*, 56.

[45] Christopher Polhem: "Letter to Ericus Benzelius, 16 July 1710", in Tafel (ed): *Documents* vol. 1, 205.

[46] Ibid.

our country will never see again."[47] Their letters were filled with mutual praise, in a manner typical of the intergenerational relationships of mechanics discussed in Chapter 2. Swedenborg recognised Polhammar's seniority and mechanical excellence, and Polhammar pointed out young Swedenborg's aptitude for mechanics. In spite of these positive comments, Swedenborg did not stay in Stiernsund to study for Polhammar. Instead, he set out on his European tour and did not return until 1714. From England, Swedenborg could present himself to his brother-in-law as even more aligned with what was expected of a mechanicus: he wrote of how he possessed an "*immoderate desire* [...] for astronomy and mechanics", and how he traversed spaces and epistemic divides of scholars and craftsmen, in order to "steal their trades, which some day will be of use to me."[48] Thus, from London, he presented himself to his brother-in-law in the ways expected of a future Swedish state official on a foreign study tour.[49]

Polhammar – Feif. An active middleman in an asymmetric relationship

It does not appear as though Swedenborg corresponded with Polhammar during his voyage. At home, Polhammar would instead initiate correspondence with Karl XII, which would shape both him and, later on, the young Swedenborg.[50] As discussed in the previous chapter, Polhammar was already established at this time as an important mechanicus with a personal *Laboratorium Mechanicum* (somewhat) integrated into the Bureau of Mines. However, in the 1710s, the mining administration came to matter less to him: instead, he established a new dominant position through direct royal contacts.[51] Following an order of Karl XII in August 1711, Wrede had changed office and had become president of a newly reformed and now independent Bureau of Commerce. It has been argued that this change should be interpreted as a removal from office, as a result of Wrede's opposition to the king's war effort. Others have pointed to the fact that Wrede could still act as a royal council [*kungligt råd*] and that his successor in the States Office and the Bureau of Accounts, Nils Stromberg (1646–1723),

---

[47] Emanuel Swedenborg: "Letter to Ericus Benzelius, 13 October 1710", in Tafel (ed): *Documents* vol. 1, 207.
[48] Emanuel Swedenborg: "Letter to Ericus Benzelius, received 30 April, 1711.", in Tafel (ed): *Documents* vol. 1, 211. Swedenborg wrote "immoderate desire" in English.
[49] On Swedenborg's visit to London, see Simon Schaffer: "Swedenborg's lunars", *Annals of science* 71:1 (2014), 4–11. On Swedish eighteenth-century study tours, see Hodacs and Nyberg: *Naturalhistoria på resande fot*, 41–4; Fors has pointed out how the foreign travels of the eighteenth century shaped the self-conception of state officials, see Fors: *The limits of matter*, 78.
[50] Fors has mentioned these letters of recommendation from the circle around Polhem to the king, but without making a closer analysis of their contents. See *The limits of matter*, 83.
[51] On Polhem and Wrede, see page 94. See also ibid., 82.

inhabited a weak position and followed the decisions of the council.[52] No matter the relative power of Wrede and Stromberg, in 1711, Polhammar established influence with the monarch that circumvented his old patron, Wrede.

Polhammar's relationship with Karl XII was channelled through the king's right-hand man, Feif. During the king's residence in Bender in the Ottoman Empire (1709–13), Feif became secretary of state [*statssekreterare*] in the field office [*fältkansli*] there.[53] On 5 October 1711, Benzelius and the members of the *Collegium Curiosorum*, by request of Polhammar himself, sent a letter of recommendation to Bender praising Polhammar. The explicit purpose of the letter was to present him as a mechanical practitioner who was useful to both the realm and the king. The letter described his aptitude as "that of a prodigy and one of the rarest in such sciences [*scientiis*] that is known in our country to this day". The members of the *Collegium* wrote that they regretted how Polhammar's work was suffering due to lack of funds during the economically difficult times. Especially, they regretted that Polhammar could only afford to work on *theoria* and lesser models, and not *praxis*, which required greater resources. In order to bridge this divide, royal support was essential. Accordingly, in the *Collegium*'s letter, we again find the tropes discussed in Chapter 2: of how a mechanicus was expected to unite the epistemic virtues of both *theoria* and *praxis*, and how their unification was a means to become a useful man committed to theocratic absolutism.[54]

At about the same time, on 9 October, Polhammar himself sent Feif a letter. Obviously, Feif had not received, let alone answered, the *Collegium*'s letter in just 4 days' time. Because of Karl XII's residence in the Ottoman Empire, the correspondence of the absolute king and his subjects circulated between the northern and south-eastern edges of the European continent. The letters of the *Collegium* and of Polhammar should instead be seen as part of a joint effort to present the mechanical practitioner to Feif and the king.[55]

---

[52] Ingegerd Hildebrand: "Falkenberg Af Sandemar, Gabriel", *SBL* 15 (Stockholm, 1956), 226; Walter Ahlström: *Arvid Horn och Karl XII 1710–1713* (Lund, 1959), 101. See also Karlsson: *Den jämlike undersåten*, 43–4.

[53] Bengt Hildebrand: "Casten Feif", *SBL* 15 (Stockholm, 1956), 512; Karlsson: *Den jämlike undersåten*, 35.

[54] "*prodigieust* af ett af de raraste, som uthi sådanna scientiis in til dagz i wårt fädernes land bekant warit"; *Collegium Curiosorum* to Casten Feif: "Copy of letter of recommendation for Polhammar", 1711-10-05, Codex Br. 31, Lindköpings stiftsbibliotek. This copy is thus a rare example of a preserved letter *to* Feif concerning Polhammar. See also Bring: "Bidrag till Christopher Polhems lefnadsteckning", 45.

[55] Polhammar also sent another letter, concerning the stipendiaries in mechanics at the Bureau of Mines on 20 November. These two letters, like all other incoming correspondence to Feif during his time in Bender, are sadly lost. Still, Feif discusses the contents of Polhammar's letters explicitly in his answers, and also mentions the dates that Polhammar sent them. Thus, it is possible to deduct at least some of their contents. In Casten Feif: "Letter to Christopher Polhammar, 1712-02-09", *Karolinska förbundets årsbok* (1911), 238–42, Feif refers to the date of Polhammar's first letter. He refers to the second letter in Casten Feif: "Letter to Christopher Polhammar, 1712-03-21", *Karolinska förbundets årsbok* (1911), 244–7. Both Feif's letters are published in Samuel E.

In many respects, Feif was the ideal recipient of these letters. As the secretary of state, he was in charge of domestic affairs and acted as a mediator between the king and officials of the bureaus in Stockholm.[56] A large number of Feif's letters from this time are scattered throughout Swedish archives. Furthermore, during the nineteenth century, some were printed in published volumes. In his correspondence, Feif comes across as a man who was following the latest French and German discussions on cameralism and policing, and who was interested in matters of œconomy and useful knowledge.[57]

Feif replied to Polhammar on 9 February. He wrote of how the king had read an essay on Polhammar's mechanical inventions, presumably attached to one of his letters, "with such delight" that he had ordered Feif "to write". The king had wanted Polhammar "to finish the machines and models mentioned in the essay as soon as possible".[58] Why did Feif present the king as interested in Polhammar's mechanical work? For a number of historians, the king's interest in Polhammar became a circular argument that proved both that Polhammar possessed mechanical genius and that Karl XII had a genuine interest in mathematics and an ability to identify aptitude for such arts.[59] Even historians who have focused on Feif have seen him as a mere conduit. Not unlike how Polhammar has been interpreted, latter-day historians have construed Feif as a forbearer of mid-eighteenth-century Swedish "utilism". For Bengt Hildebrand, Feif's correspondence "forebodes the world of ideas of the Age of Liberty," and he considers it striking "how modern they [the letters] seem to be for their time – they could just as well have been written around 1739" (i.e., the year that The Royal Swedish Academy of Sciences was founded). Still, somehow he combines this argument with an uncritical interpretation of Feif's relationship with the king. For him, it was "beyond all doubt, that [Feif] in all his correspondence [...]

Bring (ed): "Några bref från Casten Feif till Christopher Polhem", *Karolinska förbundets årsbok* (1911), 233–56.

[56] Karlsson: *Den jämlike undersåten*, 36.

[57] Letters by Feif to Tessin are printed in Gustaf Andersson and Carl Gustaf von Brinkman (eds): *Handlingar ur v. Brinkmanska archivet* vol. 1, 133–246; letters to Polhem can be found in Feif: "Några bref från Casten Feif till Christopher Polhem"; his letters to Horn are found in M Bohnstedt: "Från Bendertiden. Casten Feifs brev till Arvid Horn 1710–1712", *Personhistorisk tidskrift*, 1921. Furthermore, unpublished letters to a wide number of recipients can be found in UUB, RA, KB and in the libraries of Lund University and of the Diocese of Lindköping.

[58] "med sådant behag"; "skrifwa"; "så snart som någonsin giörligit är, nu i förstonne skal förfärdiga de på hos gående upsatt nämnde machiner och modeller"; Feif: "Letter to Polhammar, 1712-02-09".

[59] Bring: "Bidrag till Christopher Polhems lefnadsteckning", 41, 46. Similarly, Torsten Althin writes that in 1713, "Karl XII had [...] turned his attention to Polhem's mechanical genius" (in original: "Karl XII hade [...] fått sin uppmärksamhet riktad på Polhems mekaniska snille"); Torsten Althin: *Christopher Polhem och Stjernsunds manufacturverk* (Säter, 1950), 23. Also, in a more recent biographical article on Polhem, Michael Lindgren argues that Polhem and the king "found each other through a mutual technical interest." (in original: "fann varandra i ett gemensamt tekniskt intresse"); Michael Lindgren: "Christopher Polhem", *SBL* 29 (Stockholm, 1995–7), 338.

to a large degree was the direct spokesman for the king, for Karl XII himself."[60]

More recently, Åsa Karlsson has corrected this image: Feif had a much more active role as a mediator between Polhammar and the king.[61] Moreover, he was central in shaping Polhammar and the king into men with similar interests. Feif's letters were not passive or neutral descriptions of the involved correspondents. The correspondence between all the men was a forum, in which they could reshape themselves through relational performances. Feif's presentations of the king's interests can be seen as one such relational performance. By presenting Karl XII as interested in both Polhammar and mechanics, Feif could conjure forth a specific image of the monarch. He described how "His Majesty has a particular liking of mechanics" and repeatedly wrote of the king's interest in maintaining the correspondence. He wrote how he wanted Polhammar to write longer letters, which "His Majesty himself would read [...] with pleasure". Furthermore, he pointed out the haste with which the king had wished to answer Polhammar. From this, Feif argued, Polhammar should have been able to discern "what a gracious consideration His Majesty has for his work, and what he could expect in the future."[62] In a letter of 5 March, Feif once again described the king's great interest in and aptitude for mechanics, "a science to which His Majesty takes a particular liking, wherefore His Majesty is not completely unskilled in it." Presenting himself as sharing the king's interests Feif promised to use "all my diligence to persuade the whole world that among us, arts and sciences are held in appropriate respect".[63]

The relational performances found in these letters follow similar patterns to those discussed in previous chapters: they were made in a relationship of superiority and submission, between actors who were anything but peers. Nevertheless, by relating to each other, both parties shaped themselves in each other's image. The relation between the monarch and the mechanicus, formed through their correspondence, had a somewhat different power dynamic than that between the old and the young mechanical practitioner. Although Karl XII was 22 years younger than Polhammar, it is obvious that he was the senior party in the relationship. As an absolute

---

[60] "förebåda frihetstidens tankevärld"; Hildebrand: *Kungl. Svenska Vetenskapsakademien* vol. 1, 72. "hur moderna de verka för sin tid – de kunde i så måtto lika väl varit skrivna omkr. 1739"; ibid., 73. "utom allt tvivel, att han i hela sin brevväxling [...] är i hög grad ett direkt språkrör för kungen, för Carl XII själv."; ibid., 72.

[61] Karlsson: *Den jämlike undersåten*, 36.

[62] "H:s M:t har en särdeles lust til mechanicen,"; Feif: "Letter to Polhammar, 1712-02-09", 239. "H:s M:s sielf lärer läsa dem med nöije igenom", ibid., 241. "hwad nådig omtanke H:s M:t har for des arbeten, och hwad han framdeles kan hafwa at förwänta."; ibid., 242.

[63] "en wettenskap hwartil H:s M:t har särdeles lust, hwarföre Hans M:t eij heller lärer finnas så aldeles okunnig däruti."; Casten Feif: "Letter to Christopher Polhammar, 1712-03-05", *Karolinska förbundets årsbok* (1911), 243. "al min flijt at öfwertyga hela werlden, det konster och wettenskaper hos oss blifwa håldne uti tilbörligt wärde"; ibid., 243–4.

monarch, he is perhaps the most extreme case of how age and social status intersected in the patriarchal order of early modern Sweden. As an active middleman in this asymmetric relationship, Feif did not only make Polhammar in the image of a mechanicus: he also made Karl XII into a certain type of œconomically minded monarch, acting according to both mechanical and cameral principles.

Third parties. Making mechanics matter

Feif's correspondence with Polhammar was not only of concern to Feif, Polhammar and the king. A broader circle discussed the relationship between Polhammar and Karl XII, whose relation resonated in Feif's letters to other officials in Sweden. For example, Feif and Nicodemus Tessin (1654–1728) explicitly discussed royal patronage of useful arts and sciences. Tessin had made a career by offering the monarch his services as an architect and organiser of ceremonies, and, like Polhammar, his success depended on the patronage of both Karl XI and Karl XII. Compared with Polhammar, he was a much more successful client and, in 1716, he even became marshal of the realm.[64] Feif wrote to Tessin about Polhammar just weeks after receiving Polhammar's first letter. He described how "His Majesty has not known until now that he had such a man in his realm, and much laments that he has been unknowing for so many years."[65] He thus rushed to present the king's new relationship with Polhammar to his contact. Despite the king's professed previous ignorance of Polhammar, the existence of such a man in the realm was a sign of the king's interest in useful mechanical and beautiful arts. In a roundabout way, Feif wrote that, although the king did "not wish to brag about his knowledge", the Karl XII was "not unskilled in *architecture*". Because of the king's interest in this art, Feif could promise Tessin that the king would "promote beautiful and useful arts and sciences in every way. A proof thereof is *Polhammar*".[66]

In his letters, Feif made sure to highlight the king's enthusiasm for Polhammar time and time again. On 19 November, Feif again mentioned Polhammar to Tessin. Tessin had lamented how the young were uninterested in the arts. However, Feif explained how this was due to lack of encouragement, which in turn resulted from unpatriotic sentiments: "How would it

---

[64] On the relationship between Tessin and the monarchs Karl XI and Karl XII, see Snickare: *Enväldets riter*, 29–43; Snickare: "Shaping the ritual space", 133; Linda Hinners: *De fransöske handtwerkarne vid Stockholms slott 1693–1713. Yrkesroller, organisation, arbetsprocesser* (Stockholm, 2012), 46–8.
[65] "H. M:t har aldrig förr än nu vettat, at han hade en sådan man i sitt rike, och beklagar myket, at han i så många åhr har måst gå fåfäng."; Casten Feif: "Letter to Nicodemus Tessin, 1712-02-30", in Andersson & Brinkman (eds): *Handlingar* vol. 1, 146.
[66] "vil intet skryta med sine wettenskaper"; ibid. "är intet så osnäl i *Architecturen*,"; ibid., 145. "på alt sät lärer befordra vackra konster och nyttige vettenskaper, et bevijs däraf är *Påhlhammar*."; ibid., 146.

be possible for anyone to find an appetite thereof," Feif wrote, "as long as we value what strangers do considerably more". Polhammar, Feif argued, would have been praised to the skies had he been a stranger, "but because he is a Swede, he is not much esteemed for all his knowledge." But Feif continued on the positive note that because "His Majesty loves sciences, more people will doubtless commit themselves to them".[67] Likewise, Feif discussed Polhammar in letters to Arvid Horn (1664–1742), the president of the chancery [*Kanslikollegium*]. Their correspondence touched on several œconomical subjects. In this context, Feif wrote how it would please the monarch if Horn aided Polhammar who "can accomplish many useful things in our manufactories." Again, the royal patronage of Polhammar became a symbol of the king's interest in useful arts and sciences, and in turn a proof of their importance to the realm.[68]

To present the king's interest in mechanics was something more than to reveal an expression of personal opinion. As the locus of the body politic, the king's interest in mechanics was perceived as corresponding to the interest of his subjects. Therefore, the relational performances of mechanics carried out in this correspondence not only shaped Polhammar and the king, but also attributed to mechanics new meanings in relation to the political order. By writing about the relationship between the king and Polhammar, Feif could present himself as an important broker and assure other potential clients that they too could expect benevolent responses from the monarch. When Feif wrote about Polhammar to Tessin, he presented the king as in favour of work similar to Tessin's. By reading of how the king treated Polhammar, the architect Tessin could be certain of similar just treatment.

In his correspondence with third parties, Polhammar presented his relationships with Feif and the king in the same manner. In a letter of 31 May 1712 to Pehr Elvius the elder, the mathematics professor in Uppsala, Polhammar related how "Chancery Councillor [*Cancelierådet*] Feif writes that His Majesty is very much a *libhaber* [sic], or a lover, of mechanics, and knows fairly much thereof." Moreover, he told his friend of how the king "yearns, as he [Feif] writes, that he could talk to me".[69] Polhammar continued to describe to Elvius how the king cared so much for useful knowledge and its practitioners that he wished that young men would study mechanics.

---

[67] "huru är det möijeligit at någon skal få lust därtil, så länge man hoos oss tycker ålijka mera om det som främmande giöra"; "men efter han är en svensk, acktas han intet stort med al sin vettenskap."; "Hans M:t älskar vettenskaper, så lära flera lägga sig därpå"; Casten Feif: "Letter to Nicodemus Tessin, 1712-11-19", in Andersson & Brinkman (eds): *Handlingar* vol. 1, 162.

[68] "kan skaffa mycken stor nytta uti wåra manufacturer."; Casten Feif: "Letter to Arvid Horn, 1712-03-26", *Personhistorisk tidskrift* (1921), 111.

[69] "H. Cancelirådet Feif förmehler att Hanss Maij:tt ähr myket libhaber eller älskare af mechanicen, och förstår sig tembligen well der uthi."; "lengtar myket, som han skrifver, att få tala med mig"; Christopher Polhem: "Letter to Petrus Elvius, 1712-05-31", in Liljencrantz (ed): *Christopher Polhems brev*, 88.

Therefore, Polhammar informed his friend at Uppsala University that "the students [lit. subjects] of the academy, who consider themselves inclined to learn [mechanics], gain no disadvantage if they apply themselves to it early in life."[70] By writing of the king's interest in mechanics, Polhammar could present his own work as prestigious and as essential to the education of future men of the state. As in Feif's correspondence, the making of a mechanically inclined monarch was intimately related to the making of both Polhammar into a useful man and mechanics into a feasible path to a useful manhood. Thus, Polhammar recognised the connection between the mechanical interest of the monarch, and mechanics as an important part of the body politic. By staging the monarch as interested in mechanics, young men would identify it as a feasible way to fashion themselves into manly subjects.

Making Polhammar an arbiter of mechanical aptitude

The relational performances of the monarch and the mechanicus were entangled with the relationships discussed in the two previous chapters: of mechanics as a means for a boy to become a useful man, and for young men to become part of the bureaus in Stockholm. Through the stipends in mechanics, established in his *Laboratorium Mechanicum* of the Bureau of Mines in 1699, Polhammar had already attained a position where he was the authority on whether or not young men showed promise in the mathematical and mechanical arts. As seen in Chapter 3, until the 1720s, Polhammar held the authority to identify who among the young of the realm were mechanically apt. On 21 March 1712, Feif reaffirmed this position: "No one," Feif argued, "must [...] be accepted [as a stipendiary], whom [Polhammar] does not deem to have a considerable aptitude for mechanics, so that they in time can accomplish something useful". No consideration should be given to applicants' "various recommendations," Feif proposed: only their aptitude should matter. Feif gave Polhammar one suggestion as to where to search for such boys: maybe the children of master craftsmen "in the crafts that are related to mechanics" could benefit from Polhammar's education.[71]

As pointed out in Chapter 2, in an intergenerational relationship between mechanical practitioners, issues of identity were paramount: but by which qualities could someone identify an apt mechanical practitioner? In which ways did mechanical practitioners relate to other personas, such as that of the craftsman or the scholar? Feif's letter to Polhammar about the stipendiaries puts these questions in a new perspective. The didactic narra-

---

[70] "dhe subjecta som wijdh accadamien finna sig der till böjliga att lära, giöra intet illa mot sig om dhe aplicera sig der till i tijd."; Polhem: "Letter to Petrus Elvius, 1712-05-31".

[71] "måste ingen därtil antagas, som icke H:r Direct. finner hafwa det särdeles genie til mechaniquen, så at de med tiden något nyttigt kunna uträtta"; Feif: "Letter to Polhammar, 1712-03-21", 244. "hwariehanda recommendationer,"; ibid., 244–5. "uti de handtwärken som hafwa gemenskap med mechaniquen"; ibid., 245.

tives found in speeches and prefaces were not only a matter of hypothetical discussion, they were also a means for Polhammar – arbiter of the mechanical aptitude of boys by royal decree – to present systematic criteria for recognising mechanical competence. Here we can find a Swedish parallel to the French development described by Jay Smith and discussed above. By making Polhammar the single authority for the stipends, and by highlighting competence in mechanics in favour of recommendations, it seems as if Feif wished to systematise these positions in the Swedish bureaucracy. Paradoxically, through a personal relationship with the king, Polhammar was instilled with a power to sidestep the personal modality of service – based on networks of kinship and favours – in the Swedish bureaucracy. Of course, in reality, as discussed in the previous chapter, Polhammar used this authority to promote his own sons, as well as other young men who were close to him.

The establishment of Polhammar's authority, based on his perceived relationship with the monarch, can be seen in Stromberg's letters to Polhammar. As a result of having received letters from Feif, Stromberg in turn wrote to Polhammar on 6 June 1712, explaining that:

> I have recently received a letter from Bender by Chancery Councillor Feif, which among other things conveys how since His Majesty has been duly notified of [Polhammar's] great talent, and rare aptitude, for the mechanical sciences, His Majesty has shown a particular royal grace and liking of [him].[72]

Much like in a children's game of Chinese Whispers, Polhammar's initial self-presentation as a useful mechanicus in his letter to Feif in Bender had travelled from Feif to the king, back again to Feif, then to Stromberg in Stockholm, and finally back to Polhammar. What Stromberg communicated to Polhammar was his own interpretation of what Feif had told him about the king's reactions to reading Polhammar's letters. Having circulated through this chain of correspondence, Polhammar's initial self-presentation returned to him in a new form authorised by the monarch. It was no longer a subjective performance by a single man, but an officially sanctioned description of his mechanical excellence; the king, through Feif, had authorised Polhammar's self-presentation as an official truth. Carrying out the royal will, Stromberg promised Polhammar monetary and symbolic resources; he offered him his services and promised to help him financially in

---

[72] "Som iag nyl. erhållet bref från Bender af Hr Cancellie Rådet Feif, derutinnan bland annat äfwen förhähles, at sedan Hans Maj:t är worden behörigen underrättad om H:r Directeurens stora talent och ogemena genie uti Mechaniska wettenskaper, har Hans Kongl. Maj:t låtit förspörja en serdeles Konglig ynnest och benägenhet för H:r Directeuren,"; Nils Stromberg: "Letter to Christopher Polhammar, concerning the King's recommendation of Polhammar", 1712-06-03, I p:23 1, 64, KB.

order to encourage "the completion and maintenance of his ingenious speculations and his useful machines".[73]

Polhammar's authority as a mechanicus was established by repetitious performances of self-presentation, which were passed on and amplified through a chain of correspondence. Simultaneously, the chain reinforced the role of mechanics as an identificatory practice of a useful subject. This was not the first time Polhammar had fashioned himself into such an ideal subject, nor would it be the last.[74] Moreover, Polhammar was not alone in using the correspondence to acquire royal authority. All parts of the chain – be they monarchs, secretaries or civil servants – used the relationship between Polhammar and the king to do the same. Interestingly, the correspondence from Polhammar to Feif and back again took place in a relatively short time between February and November 1712. It seems as if a perceived relationship between Polhammar and the king was in the interest of many actors at this specific time. For Polhammar, it was a means of establishing authority for himself as well as for the mechanical arts; for Feif and members of the civil administration in Stockholm, it was a last ditch attempt, after the military failure of the battle of Poltava in 1709, to re-form Karl XII into a cameral monarch with an interest in manufactories.

A sudden silence

This circulation of correspondence, between two corners of Europe, was carried out over the improvised and fragile infrastructure that the king used to maintain contact with the administration in Stockholm while abroad. On 13 December 1712, Feif sent one last letter to Polhammar before the correspondence was put to a halt by the events of the Skirmish at Bender [*Kalabaliken i Bender*] in February 1713. Karl XII had outstayed the welcome of his Ottoman hosts, and was now captured and forced to return to Sweden.

However, in the summer of 1713, Feif re-established his correspondence with Tessin. On 22 June, he wrote a long letter to his friend. In this letter, Feif again touched on œconomy and policing, and expressed his wish to procure works by Wilhelm von Schröder and Marc-René d'Argenson. On 13 July, Feif continued his discussion on cameral matters in yet another letter. He admitted that he wished Tessin's letters (just like Polhammar's) to be "four times longer than they are, because they very much please His

---

[73] "sinrijka Speculationer och nyttige Konst-Machiners förfärdigande och upprätthållande"; ibid. Polhammar acquired similar help by decree from the government's office to the council, the clearest example perhaps being tax and customs relief for his manufactory in Stiernsund. See e.g.: Karl XII: "Copy of letter to the council on new privileges for Stiernsund", 1712-03-13, I p:23 1, 59, KB.

[74] It would be possible to make a similar analysis of Polhammar's relationship with Karl XI and Fabian Wrede, when he received stipends for travelling abroad and the *Laboratorium Mechanicum* of the Bureau of Mines. Likewise, the refashioning of the then ennobled Polhem into a typical man of the "Age of Liberty" would reveal another remaking of the mechanical practitioner.

Majesty. They are read more than one time." He also described his library of books on œconomy and policing, and argued that "our œconomy and police need a considerable library, if they are ever to be put in a good condition." Feif even suggested that Tessin should consider the establishment of a Swedish "*Academie de* [sic] *Sciences*", which Tessin should preside over.[75]

Although Feif continued his correspondence about economics with Tessin in 1713, there are no preserved letters from Feif to Polhammar from that year. It was not until the autumn of 1714 that Polhammar again received letters from the chain of correspondents that he and the king were part of. On 6 October 1714, Stromberg communicated to Polhammar the king's decision to award him the position of extraordinary assessor in the new Bureau of Commerce.[76] On 20 December, Feif wrote to Polhammar again, ordering that he should prepare to travel to the king in Stralsund as soon as the ice broke.[77] This voyage never came to pass. Instead, Polhammar, Feif and the king would not meet until the king returned to Sweden. Accidentally, this coincided with Swedenborg returning from his European tour.

Stiernsund revisited. The affection of Polhammar and Swedenborg

When Swedenborg returned to Sweden, he re-established his relationship with Polhammar. In a letter to Benzelius of 8 September 1714, he expressed

a very great desire to return home to Sweden, and to take in hand all Polhammar's inventions, make drawings, and furnish descriptions of them, and also to test them by physics, mechanics, hydrostatics, and hydraulics, and likewise by the algebraic calculus.[78]

Between 1716 and 1718, and with Benzelius' assistance, Swedenborg started to publish Polhammar's inventions and other curious works in the journal *Daedalus Hyperboreus* under the auspices of the *Collegium Curiosorum* in Uppsala. Through this work, Swedenborg would become increasingly connected to Polhammar and consequently, by proxy, to Karl XII. Together, the relationships of these men would become a vortex through which all those involved were shaped into useful men of the absolutist regime.

Writing to Swedenborg on 7 December 1715, Polhammar praised the younger man's work on publishing his inventions in *Daedalus Hyperboreus*. To

---

[75] "fyra gånger så långa som de äre, medan de myket förnöija Hans M:t de läsas mehr än en gång igenom,"; "vår *œconomie* och vår *police* behöfver anseenlige *bibliothek*, om de skola komma i god stand."; Casten Feif: "Letter to Nicodemus Tessin, 1713-07-17", in Andersson & Brinkman (eds): *Handlingar* vol. 1, 180.

[76] Nils Stromberg: "Letter to Christopher Polhem", 1714-10-06, I p:23 1, 82, KB.

[77] Casten Feif: "Letter to Christopher Polhem", 1714-12-20, I p:23 1, 86, KB.

[78] Emanuel Swedenborg: "Letter to Ericus Benzelius, received 8 September 1714", in Tafel (ed): *Documents* vol. 1, 232. This section is to a high degree based on letters found in Tafel's collection of English translations of letters related to Swedenborg. Therefore, I do not present the original quotes here. References to the original sources can instead be found in Tafel (1875).

him, Swedenborg was "a ready mathematician, and well qualified for doing this and similar achievements." Moreover, he was pleased with the praise he had received from Swedenborg in a proposed preface of the work, although he wished the young man to tone it down "so that the sense of delicacy may not be offended thereby".[79] Three days later, Polhammar praised Swedenborg in a letter sent to Benzelius:

> Young [Swedenborg] is a ready mathematician and possesses much aptitude for the mechanical sciences; and if he continues as he has begun, he will, in course of time, be able to be of greater use to the King and to his country in this [capacity] than in anything else.[80]

As before Swedenborg left on his tour, the two men were tied together through relational performances of praise. On 19 December 1715, Polhammar invited Swedenborg to Stiernsund, stating that he would "experience both pleasure and delight in discussing upon [my mechanical designs] with one who is interested in them; for otherwise it would be like loving some one [sic] by whom you are not loved in return."[81] In September the next year, Polhammar yet again invited Swedenborg to his home in Stiernsund:

> Your arrival in Stiernsund will be most agreeable to me, and if my experience can be of any use to you, I will give it with so much the greater pleasure, as the fruit of it will be of use to the public and will accrue to my own honour.[82]

The letters present a loving relationship between Polhammar and the much younger Swedenborg, based on both emotional dependence and mutual benefits. Like the relationship between Polhammar and the king, it was an asymmetric relationship between two actors of different positions. In this case, the relationship between Polhammar and Swedenborg followed the imagined intergenerational relationships of mechanics, whereby Polhammar adopted the superior role of the older man and Swedenborg the subjugated role prescribed to the young. Still, by publishing Polhammar's inventions, Swedenborg did not only act as a passive apprentice being shaped by his master, but also presented Polhammar as a Northern Archimedes, or Daedalus, to a wider European audience of scholars as well as to the king. Thus, again we find a relationship of superiority and submission in which both parties gained authority and were shaped in relation to each other.

---

[79] Christopher Polhem: "Letter to Swedenborg, 7 December 1715", in Tafel (ed): *Documents* vol. 1, 242.

[80] Christopher Polhem: "Letter to Ericus Benzelius, 10 December 1715", in Tafel (ed): *Documents* vol. 1, 243–5.

[81] Christopher Polhem: "Letter to Swedenborg, 19 December 1715", in Tafel (ed): *Documents* vol. 1, 246.

[82] Christopher Polhem: "Letter to Swedenborg, 5 September 1716", in Tafel (ed): *Documents* vol. 1, 271.

Daedalus and the sovereign. The conflation of two relationships

Soon after Swedenborg's return to Sweden, the king also re-appeared in the realm. By the early autumn of 1716, Karl XII had settled in the university town of Lund in the southernmost part of the country. The king stayed in Lund until 1718, and while residing there he took an interest in the work of the academy.[83] He visited a mathematical lecture given by the professor in astronomy and mathematics, Conrad Quensel, and Johan Jacob Döbelius' defence of a dissertation on the nature of sense perception. Furthermore, he frequented Andreas Rydelius' lectures in philosophy.[84] Like the correspondence studied earlier, the king's stay in Lund provided ample opportunities for various actors to reinvent themselves as well as the king.

As Swedenborg was set to travel to Stiernsund in September 1716, Polhammar received command from the king to join him in Lund. Polhammar invited Swedenborg to accompany him and, in preparation for their voyage, Swedenborg made a stop at Benzelius' in Uppsala to prepare a special bound edition of the four issues of *Daedalus Hyperboreus* explicitly written for a royal audience. This edition of the *Daedalus* was a tailor-made gift for Karl XII, printed on high-quality paper, and it included a dedication to the king written by Swedenborg just for this occasion.[85] Swedenborg commenced his dedication with a poem:

> Lo Daedalus did mount the winds, and from on high
> > Did scorn the snares King Minos laid on earth.
> So mount the winds, my Daedalus, by thine own art
> > And scorn the snares the common herd shall lay.[86]

In his poem, Swedenborg related to the ancient myth of Icarus and his father Daedalus, the mechanicus and namesake of Swedenborg's journal. In the myth, hoping to escape the island of Crete, Daedalus explores unknown arts in order to fabricate wings for his son and himself. Taking off into the sky, Icarus and Daedalus are seen by common men, as described in an English renaissance translation of Ovid's version of the myth:

> The fishermen Then standing angling by the Sea, and shepeherdes leaning then | On sheepehookes, and the Ploughmen on the handles of their Plough, | Beholding them, amazed were: and thought that they that through | The Aire could flie were Gods. And now did on their left side stand.[87]

---

[83] Hatton: *Charles XII of Sweden*, 428–41; Liljegren: *Karl XII i Lund*, 57–69.

[84] Hatton: *Charles XII of Sweden*, 430. For a discussion of Rydelius' philosophy in relation to the absolutist theocracy, see Chapter 2.

[85] Alfred Acton (ed): *The letters and memorials of Emanuel Swedenborg* vol. 1, 1709–48 (Bryn AthynPA, 1948), 121.

[86] Ibid., 122.

[87] Ovid: *The XV bookes of P. Ouidius Naso, entytuled Metamorphosis, translated oute of Latin into English meeter, by Arthur Golding Gentleman, a worke very pleasaunt and delectable* (London, 1567), 99.

Sawday discusses two roles given to invention in the Icarus myth by renaissance culture. First, the myth was read as a tragedy of Icarus' fall from the sky. Consequently, in the renaissance, Daedalus was not only a positive character. For example, for Bacon, Daedalus was a symbol of mechanical skill used for a negative purpose.[88] At the same time, the myth was the story of how a certain kind of man could bend nature to his own purpose.[89] In his poem, Swedenborg picked up on this second role of the Icarus myth, but he focused less on invention as a process of bending nature than as a process of shaping the inventor himself. Through his "own art", Swedenborg's Daedalus would be elevated to a position from which he could "scorn the snares the common herd shall lay". Knowing the intended reader for his work, the lines of verse presented mechanics as a technique not only for Swedenborg and Polhammar, but also for the king, to reaffirm their authority elevated above other subjects of the realm.

Swedenborg's ensuing dedication, written in the expected submissive style, is a stark contrast to the men flying in the winds. Here he stated that he had wished "to come forward with some small mathematical investigations and observations, and lay them down in deepest submission at your Majesty's feet". This he dared to do because of the monarch's recognised interest in:

> literary art in general and, in particular, in *studia mathematica*, a signal proof whereof is the fact that your Royal Majesty has ever regarded with grace the designs and machines which Assessor Pålheimer [sic] has already set up for the service and use of your Royal Majesty and his Kingdom.[90]

Finally, Swedenborg concluded the preface by linking the presented work, as well as the king's reaction to it, to the conditions for mechanical practitioners as a whole: "If this work wins your Majesty's grace, it will certainly rouse up many other men, in submissiveness to lay bare their thoughts, and to offer them for your Royal Majesty's gracious pleasure."[91]

Acton argues that the preface "was clearly designed to enlist the King's support".[92] But apart from this obvious observation, what is striking is *how* this was done. Swedenborg's text intertwined the relational performances of Swedenborg–Polhammar and Polhammar–Karl XII into a coherent whole. The king's interest in mathematical and mechanical arts, and the recognition of Polhammar's work were, using the circular logic uncritically related by older historians, proof of each other. In the dedication, Swedenborg grafted himself onto this relationship as the junior partner in his and Polhammar's intergenerational relationship. The combination of these two relationships

---

[88] Sawday: *Engines of the imagination*, 215.
[89] Ibid., 24–30.
[90] Acton (ed): *The letters and memorials of Emanuel Swedenborg* vol. 1, 122.
[91] Ibid., 122.
[92] Ibid.

made it possible for Swedenborg to present all three men as useful and virtuous through his mathematical work.

On 6 December, after Polhammar and Swedenborg had arrived in Lund, Polhammar submitted another "humble memorial" recommending Swedenborg to the king. In this text, the presentations of Polhammar, Swedenborg and the king, and the virtues of mechanics once again merged. First, Polhammar noted how "your Majesty's grace and delight in the mechanical sciences has become so manifest". He also pointed out how the king knew that "mechanics is a study which demands much labor and brainwork". Furthermore, he argued that mechanics had until now been held in low esteem. It had been seen "as the art of a common workman", in spite of the fact that it required "the best subjects and the quickest talents that can be found in nature". The only way to rectify this injustice would be if "subjects who are skilled in this science be regarded with no less honor than others whose studies are merely mediocre."[93]

Once again, the monarch's interest in mechanics was supposed to correspond to the role of mechanics in the Swedish realm, and again Polhammar picked up on this connection. He presented the young Swedenborg as a possibility for the king to give mechanics the recognition it deserved: "At this time I know of no one who seems to have a greater bent for mechanics than Herr Emanuel [Swedenborg]." Nevertheless, Polhammar complained, Swedenborg did not study mechanics. Polhammar explained why: "that he applies himself to other studies, is caused by the small regard in which, according to former custom, mechanics is held." [94] Therefore, Polhammar asked whether it would not be "useful to grant some prerogative of honor to one who has a natural bent for mechanics".[95]

In a letter to Benzelius, from late December 1716, Swedenborg related the king's positive reaction to Polhammar's request, while he reinforced the image of the king's interest in mechanics: "Since His Majesty graciously looked at my *Dædalus*," Swedenborg wrote, the king had "advanced me to the post of an extraordinary assessor in the [Bureau] of Mines; yet in such a way, that I should for some time attend the Councillor of Commerce, Polheimer [sic]".[96] It pleased Swedenborg that the king "pronounced so favourable and gracious a judgment respecting me, and himself defended me against those who thought the worst of me; and that he has since promised me his further favour and protection". Moreover, he noted how the "*Daedalus* has enjoyed the favour of lying these three weeks upon His Majesty's table, and has furnished matter for many discussions and questions". It had

[93] Ibid., 125.
[94] Ibid.
[95] Ibid., 125–6.
[96] Emanuel Swedenborg: "Letter to Benzelius, December 1716", in Tafel (ed): *Documents* vol. 1, 273–4.

even, Swedenborg noted, "been shown by His Majesty to many persons."[97] Lying on the king's table, his book became a manifestation of the relationships, formed around mechanics, between the monarch and the mechanici. Swedenborg's letter was a performance on several levels. First, it contained a description of how his *Daedalus* was performed in Lund, giving him a position as well as prestige. This description, in turn, was a performance in its own right, by which Swedenborg presented himself as a mechanicus by royal decree and fashioned the king into a monarch with an interest in mathematics. As in the case of Polhammar, the asymmetric relationship between Swedenborg and the king thus involved the exchange of mutual favours as well as mutual presentations of selves. Interestingly, not only did Swedenborg graft himself onto Polhammar through his presentations to the king, the royal gifts Swedenborg received tied him even closer to the older mechanicus.

On 10 December, the king wrote to the Bureau of Mines that he was pleased to "advance *Em: Sw.* to be *Extraord: Ass.*," but that he should "accompany Councillor of Commerce Polhammar and be his assistant in instituting his constructions and inventions".[98] While binding Polhammar and Swedenborg to his royal persona, the king bound the two mechanical practitioners together as master and assistant. In the process, the king raised the two mechanical practitioners to the upper echelons of the Swedish civil bureaucracy, and consolidated mechanics as an identificatory practice of a useful manly subject. Suitably, it was at this time, in December 1716, that Christopher Polhammar was ennobled and became Christopher Polhem.[99]

The limits of an absolute monarch

How was the king shaped by these relationships? As pointed out earlier, in the chain of letters in 1712, the king was already presented as a man with a taste for mechanics. Polhammar argued that, as a consequence of this alleged royal interest, the role of mechanics in the realm had changed too. In the ensuing events of 1716, we see that the monarch's interest in mechanics is already taken for granted in the performances of Polhammar and Swedenborg: the two men shaped themselves in relation to an imagined mechanicus monarch. Furthermore, the performances of Polhammar and Karl XII reinforced the monarch as a mechanical *Liebhaber*, or a lover of mechanics. Mechanics seems to have worked as a symbol through which Karl XII and his administration could present themselves as caring for the œconomy of the realm during an economic and military crisis. Still, there

---

[97] Ibid., 274.
[98] Acton (ed): *The letters and memorials of Emanuel Swedenborg* vol. 1, 136–7.
[99] For a discussion of Polhammar's ennoblement, see Bring: "Bidrag till Christopher Polhems lefnadsteckning", 49.

were boundaries to how the king could engage with mechanics and mathematics; his interest could not transgress the asymmetrical logic of a relationship between sovereign and subject. The emotional knots that bound the monarch and the mechanici together were thus not unproblematic. By being related to each other, they risked acting in ways that were unsuitable for men of their respective position.

The limits of the king's performance in relation to knowledge makers are seen during some events in the autumn of 1717. At this time, Karl XII composed his *Antropologia physica*, fourteen "theses" that many of his biographers have taken to testify to the king's interest in the philosophical theories of his day. *Antropologia physica* discussed how human beings react to stimuli, and follow the mainstream of enlightenment thinking on these matters. Karl XII started to work on his theses after having discussions with David Hein, a diplomat from Hesse and a former student of the Halle pietist Christian Thomasius. He had Feif translate them into German, and wished him to send them to Hein for comments. However, Feif advised the king against it: he argued that the king should leave it to the philosophers to take sides in philosophical controversies. If Hein were to show the king's theses to Thomasius, the philosopher might turn them into print with his own commentaries. In turn, this would make it possible for Thomasius to refute the absolute king like he would any other fellow philosopher, which, argued Feif, would be negative to the prestige of the Swedish realm.[100]

Possibly, Feif had religious motives for discouraging the king from engaging with the pietist philosopher.[101] In the light of the conflict between pietism and Lutheran orthodoxy, discussed in Chapter 2, this is not unlikely. But there also existed political reasons for the king to keep a distance from Thomasius. Having befriended and defended Johann Reinhold Patkul – a Livonian who had argued for the independence of this province of the Swedish empire and who had been convicted of high treason in his absence – Thomasius was not any philosopher, but a chess piece in the foreign policy of the king.[102] Irrespective of his motives, here, as in his correspondence,

---

[100] Hatton: *Charles XII of Sweden*, 430–1.

[101] Ibid., 431.

[102] In 1701, Thomasius had published a defence of Patkul, Christian Thomasius: *Gründliche iedoch bescheidene Deduction der Unschuld Hn. Joh. Reinhold von Patkul* (Leipzig, 1701). Still, Patkul was executed in 1707 after having come in the hands of Karl XII. His death was debated in a large number of writings all over Europe, well into the middle of the eighteenth century, and must hence still have been a matter of concern for the Swedish king in 1718. See for example: Lorentz Hagen: *Das schemertzliche doch seelige Ende, des welt-bekandten Joh. Reinhold Patkuls* (Cologne, 1714); Lorentz Hagen: *Een kort verhaal, wegens de verschrikkelyke dood, van den vermaarden heer Johann Reinold van Patkul, generaal van zyn koninklyke majesteyt van Zweeden* (Amsterdam, 1718); Michael Ranft and Johann Samuel Heinsius: *Die merkwürdige Lebensgeschichte derer vier berühmten schwedischen Feldmarschalle, Grafen Rehnschild, Steenbock, Meyerfeld und Dücker nebst dem angefügten merkwürdigen Leben und jämmerlichen Ende des bekannten Generals Johann Reinhold Patkuls. Zur Erleuterung vieler wichtigen Umstände der Geschichte Königs Caroli XII von Schweden, ans Licht gestället von einem Liebhaber der neuesten Historie* (Leipzig, 1753); Lorentz Hagen: *Anecdotes concerning the famous John Reinhold Patkul. Or, an authentic*

Feif was concerned with how the king presented himself to his subjects as well as to the rest of Europe. In this case, he cared about what the king should *not* be seen as doing. Feif's actions show how he was aware of the fine line between presenting the king as a *Liebhaber* and as someone who himself had pretentions to participate in mechanical or philosophical discussions. They also reinforce the point that patronage was not only a matter of appearances for the client, but maybe even more so for the patron.

This again illustrates the importance of maintaining an asymmetric and hierarchical relationship between the monarch and mechanici or philosophers. By intervening directly in philosophical discourse, the king would risk becoming subordinated to a specific philosophical school, and by extension, in this case, to the philosopher Thomasius. Such an action would break the asymmetric logic of these relationships of submission and superiority, and could thus undermine the king's position as an absolute authority. Feif's advice to him is similar to the discussions of how mechanically apt boys should interact with craftsmen, discussed in Chapter 2: he should aim to become identified as a specific cameral kind of monarch by entering into relationships with mechanical practitioners, but he should avoid becoming one of them, because this involved acts of submission unfitting for a man of his standing.[103]

Feif's advice highlights the complex mutual self-fashioning of royal patronage. In the case of Polhammar's patronage of Swedenborg, there existed a constant presupposition that the intergenerational relationship facilitated the younger of the two to fashion himself into the image of the older. But the relational performances of the mechanicus and the monarch were somewhat different. They were never made into each other's image, but still they were formed in relation to each other. Karl XII, Polhammar and Swedenborg never became similar; still, they became part of each other, having interlaced their persons through a stylised repetition of acts mediated in letters. When these two asymmetric relationships conflated, mechanics and mathematics were reconsolidated as identificatory practices of a pious subject.

---

relation of what passed betwixt him and his confessor, the night before and at his execution. Translated from the original manuscript, never yet printed (London, 1761); Anonymous: Breve ragguaglio della vita e della morte del conte Giovanni Patkul, nobile di Livonia, tradotto dal idioma inglese per illustrazione della vita di Pietro il Grande ... e di Carlo duodecimo. (Lugano, 1761). On Patkul and his writings on Livonia, see Pärtel Piirimäe: "The pen is a mighty sword. Johann Reinhold Patkul's polemical writings", Die baltischen Länder und der Norden. Festschrift für Helmut Piirimäe zum 75. Geburtstag (Tartu, 2005), 314–41. On Karl XII, Thomasius and Patkul, see Brunius: Andreas Rydelius och hans filosofi, 14–15.

[103] Compare to the discussion on the mechanicus and the craftsman on page 72–76.

The death of a monarch – reinventing the mechanicus

The last sentence of the previous section could have been the end of a narrative, had not the basis for the royal authority given Polhem, now ennobled, and Swedenborg soon been swept away. On the morning of 30 November 1718, while he was laying siege to Fredrikshald in Norway, Karl XII died from a stray bullet that penetrated his head. In the aftermath of his death, the role of the monarch in the Swedish political order changed. Over the following years, Sweden's government would change too, and with it the expectations of how to perform mechanics.[104]

The king's death affected Polhem profoundly. In the French edition of his travel journal, the travelling French protestant and naturalised Englishman Aubry de la Motraye discussed Karl XII's support for Polhem. Motraye had lived with the king in the Ottoman Empire, and had followed him on his northward journey to Sweden. There, he had lived with the king in Lund, and he had travelled to Uppsala, to Polhammar's Stiernsund and finally to Lappland, the unsettled lands at Sweden's northern periphery. Motraye wrote of how Karl XII had ennobled Polhammar, and thus made him into Christopher Polhem, and that Polhem had planned several mechanical projects when Motraye was in Sweden. But Motraye did "not think that the changes in Sweden since that time have allowed him to continue with them."[105] As pointed out by Motraye, these men, who had fashioned themselves into trustworthy mechanici by becoming entangled with the former king, had partly lost these benefits in the new political system.[106] Not only did their networks of patronage lose power, the authority that they had cultivated in relation to the absolute monarch diminished.

The symbol of Polhem as a Swedish Daedalus, made in relation to the absolute monarch, did not survive for much more than a year. But it did not stay dead long: the authority given to Polhem during the Swedish absolutism was to be revived in the late 1730s. Then, Jöran Andersson Nordberg rehabilitated Karl XII in an officially sanctioned biography, which included a letter by Swedenborg in which he narrated his mathematical encounters with the late king.[107] At the same time, the networks around the newly founded Royal Swedish Academy of Sciences remade Polhem into a

---

[104] A description of the events after the death of Karl XII can be found in Lennart Thanner: *Revolutionen i Sverige efter Karl XII:s död. Den inrepolitiska maktkampen under tidigare delen av Ulrika Eleonora d.y:s regering* (Uppsala, 1953).

[105] "je ne crois pas que les changemens arrivez en *Suede* depuis lui ayent permis de le continuer."; Aubry de La Motraye: *Voyages du sr. A. de La Motraye, en Europe, Asie & Afrique; où l'on trouve une grande variété de recherches geographiques, historiques & politiques* vol. 2 (The Hague, 1727), 306.

[106] Bring also picks up on this changed attitude towards Polhem, pointing out that "when Karl XII fell for the bullet in Fredrikshald, Polhem almost disappeared from the stage for a while." (in original: "när Karl XII föll för kulan vid Fredrikshall, nästan försvann Polhem för en tid från skådeplatsen,"); Bring: "Bidrag till Christopher Polhems lefnadsteckning", 57.

[107] Nordberg: *Konung Carl den XII:tes historia*, 599–600.

figurehead of their programme of useful knowledge, laying the foundation for the historiography of Polhem as a man of the so-called "Age of Liberty". The groundwork for the heroic narratives of Polhem of the nineteenth century was thus ironically lain down by the performances by which first Polhammar, later Polhem, remade himself in relation to the shifting political orders of his time. In the early 1700s, Polhammar hardly exhibited the visionary power of imagining an industrial future. At this time, Polhammar was first a man of his time and then struggling to keep up with it.

Conclusions. Interrelated personas

In early modern Europe, the monarch and the mechanicus were interrelated on a structural level, in ways that ascribed the two personas meanings that could be pieced together and actualised in actors' performances. Through these mechanistic frameworks, by which society and nature could be made intelligible, patronage of mechanics became a means of staging both monarchs and mechanics. Relationships between monarchs and mechanical practitioners were thus never only personal: they were also manifestations of a political order. Through relational performances within this order, the king was staged as an enlightened and perceptive sovereign and the mechanical practitioner as a mechanicus, a pious subject diligently offering his mechanical services to the *publicum*.

In this chapter, I have focused on two asymmetric relationships, which were conduits of relational performances that shaped both parts of each relationship. First, I have analysed the relationship between Polhammar and the young Swedenborg, a clear example of the intergenerational relationships of mechanics discussed in Chapter 2. This relationship was based on the submission of the younger party. However, Swedenborg was not only passive in his relationship to Polhammar: his act of submission did not only deprive him of agency, but also enabled him to take action. Second, I have discussed the relationship between Polhammar and Karl XII, which was actively mediated by the king's secretary Casten Feif. This relationship was based on submission and superiority, and here too acts of submission enabled action. In Feif's letters, we find performances that staged Polhammar as a mechanicus, and the king as a cameral monarch interested in œconomy and mechanics. Unlike the relationship between Polhammar and Swedenborg, this relationship was never personal. The staging of these two men was made through a chain of correspondents, with Feif in the centre. In the end, it is not even certain if Karl XII was personally involved in his supposed relationship with Polhammar. Nonetheless, Feif's staging of these two men had ripple effects throughout the Swedish realm. Not only did the staging of the king present mechanics as an identificatory practice of a pious subject, suitable for able young men to engage with; moreover, by being tied to the king, Polhammar reconsolidated his authority to identify mechanical aptitude in young men. This case is thus a concrete example of how early modern state power and knowledge making relied on, and legitimised, each other.[108] At the same time, there were limits to how one could act within these relationships. Only by acting in accordance with their asymmetric logic could the junior and senior parties maintain the relationship in spite of inhabiting vastly different positions in the absolutist order.

When Swedenborg dedicated the *Daedalus* to Karl XII, these two relationships conflated. The king became the patron not only of one mechanical practitioner, but also of an intergenerational relationship of mechanics. This triangular structure of patronage manifested not only the proper relationship between subjects and sovereigns, but also between the young and the old. As such, the conflated relationship between these men became a symbol for how mechanics was a means of becoming a pious subject in the absolutist order. Although strong, this symbol of pious mechanics would not outlive the death of Karl XII in 1718. After the king's demise, mechanical practitioners needed to find new means of aligning their techniques with the political order. In the next and final chapter, we turn to one such attempt by the previously mentioned Anders Gabriel Duhre.

---

[108] As such, the correspondence studied in this chapter supports Peter Dear's findings in Dear: "Mysteries of state, mysteries of nature".

5. A mechanicus on the kneeler. Duhre's balancing act of vice and virtue

Anders Gabriel Duhre was perhaps the most prolific author of mechanical projects of the 1720s. After a career in the civil and military administration in the 1710s, in a series of works from 1722 he proposed to transform society by founding, what he called, a *Laboratorium mathematico—œconomicum*. Like the already discussed *Laboratorium mechanicum* of the Bureau of Mines, Duhre's *Laboratorium* would be primarily an educational institution. He envisioned it to spearhead reform of mining and agriculture: as its name implied, it would revitalise the realm by raising men with a mathematical and œconomical approach to work.[1]

By the early 1720s, the political order of the Swedish state changed, and with it the relevant audiences of mechanical performances. In Chapter 4, I showed how the monarch and the mechanicus were interrelated by asymmetric relationships of patronage. However, in the constitutional monarchy of the 1720s, royal patronage increasingly needed to be complemented with support in parliament and in the bureaus. Thus, in comparison to the earlier regime – where personal relationships with the king and his government office were the basis for a mechanicus' authority – mechanical practitioners of the 1720s needed to relate to a broader group of actors. Such conditions also required new forms of self-presentation, made in relation to heterogeneous audiences. Mechanical practitioners presented themselves through printed publications, and proposals to parliament, and by participating in the communities of the state administration. The aim of this chapter is to understand how early mechanical practitioners performed in this much more complex web of relationships.

Duhre's publications were initially well received by their intended audience: the recently empowered Swedish parliament and the state administration in Stockholm. Soon the state awarded him the lease of the royal estate of Ultuna [*Ultuna kungsladugård*], near Uppsala. The income from the estate was supposed to support a permanent *Laboratorium*, where mathematics would be joined together with "useful crafts". But Duhre's *Laboratorium* would not last. Soon Duhre became entangled in controversies and disputes, which undermined his work. In 1731, he lost the estate to another

---

[1] Duhre: *Wälmenta tanckar*, 4–5.

tenant, the new county governor [*landshövding*] of Uppsala, Johan Brauner (1668–1743). In this chapter, I follow Duhre from his early days as a student, through his rise as a well-esteemed mechanicus and mathematicus, to the eventual suspension of the *Laboratorium* (in 1731) and to his death in 1739. Because of the vast amount of material concerning the *Laboratorium* – in the form of three printed publications, two extensive bound folios of handwritings left in the parliamentary archive and numerous other documents – Duhre is a unique case for understanding how actors of the early constitutional monarchy could present themselves as faithful subjects through mathematics and mechanics. Despite this fact, Duhre's *Laboratorium* has been little studied by historians. Perhaps his *Laboratorium* is too contradictory – embedded in the agricultural society and cameralism of the early modern Swedish state – to find a place in twentieth-century narratives of early modern mechanics as the midwife of industrialism. The clear failure of Duhre's project makes it virtually impossible to use him as a starting point for a heroic narrative of a man ahead of his time. Consequently, he has often been characterised as the opposite of Polhem: as a dreamer and a visionary but a useless project maker. Also, Duhre's personal qualities – described as those of a competent mathematician who found it difficult to speak to or interact with his fellow men – have been used to explain the failure of his *Laboratorium*.[2]

Here I seek understanding that goes beyond individual psychology. Duhre cannot be described as an introvert and shy mathematician: he was also valued as a teacher, and he competently presented himself as relevant to the rising parliamentary system of the early 1720s. He walked on a razor's edge between successfully performing as a mechanicus and ending up as the unsociable dreamer twentieth-century historians made him out to be. And although, at one time, he might have been considered the embodiment of a pious mechanicus, he would soon fall from grace by failing to live up to his

---

[2] The greater part of the studies that discuss Duhre were written in the first half of the twentieth century. Most of these studies are either dismissive of Duhre's work, describing it as the product of a dreamer, or only mention him in passing. Duhre and his work are discussed in: Ewert Wrangel: *Frihetstidens odlingshistoria ur litteraturens häfder 1718–1733* (Lund, 1895), 104–5; Georg Schauman: *Studier i frihetstidens nationalekonomiska litteratur. Idéer och strömningar 1718–1740* (Helsinki, 1910), 112–16; Per Magnus Hebbe: "Anders Gabriel Duhres 'Laboratorium mathematico-oeconomicum'. Ett bidrag till Ultunas äldre historia", *Kungl. Landtbruks-akademiens handlingar och tidskrift* 72 (1933), 576–94; Hildebrand: *Kungl. Svenska Vetenskapsakademien* vol. 1, 175–87; Bengt Hildebrand: "Anders Gabriel Duhre", *SBL* 11 (Stockholm, 1945), 506; Anders Grape: *Något om Anders Gabriel Duhre och en honom ägnad latinsk dikt* (Stockholm, 1949); Anders Grape: *Ihreska handskriftssamlingen i Uppsala universitets bibliotek* vol. 2 (Uppsala, 1949), 285–8; Tore Frängsmyr has discussed Duhre in relation to Wolffianism in Sweden in *Wolffianismens genombrott i Uppsala*, 64–6; More recently, Duhre and his laboratorium have been mentioned briefly in Lindqvist: *Technology on trial*, 99, 252; Olov Amelin: *Medaljens baksida. Instrumentmakaren Daniel Ekström och hans efterföljare i 1700-talets Sverige* (Uppsala, 1999), 47, 190; Staffan Rodhe has given what could be called an internalist history of Duhre's mathematics in *Matematikens utveckling i Sverige fram till 1731* (Uppsala, 2002), 50–84.

audiences' expectations. Using the case of Duhre, I thus argue that mechanics was not a one-time show, but that mechanical practitioners continuously needed to enact the norms and skills that were expected of them. Furthermore, when mechanical practitioners succeeded, their performances increased in complexity. Because of this rising complexity, initial success, accomplished by mechanical practitioners' grand promises, could ironically set the stage for eventual failure.

Becoming a good mechanicus

The father of Anders Gabriel Duhre, Gabriel Andersson Duhre (deceased 1726), was a student of the nation of Södermanland, one of the regionally divided student societies of Uppsala University. After his studies, he became a tutor [præceptor] of the royal pageboys between 1672 and 1691. As a reward for his work at court, in 1676 Gabriel received the knight's fee [Rusthåll] of Gnista in the parish of Vaksala, east of Uppsala. At Gnista, he lived with his first wife and the couple had two sons: Carl Gabriel Duhre (1680–1753) and Anders Gabriel Duhre (c.1681–1739). In March 1694, Gabriel's wife died and was buried in the churchyard of the local parish church. Exactly 1 year after her funeral – following the expected pattern of how a boy on the verge of becoming a man was to substitute experienced male tutors for his mother – the boys left home. Following in their father's footsteps, on 7 March 1695, Anders and Carl matriculated at Uppsala University.[3]

While the Duhre family was not of noble heritage, it still belonged to the upper echelons of the Uppsala community. As manifested by the knight's fee of Gnista, the family gained its means of subsistence through royal and state patronage. Like their father before them, Anders and Carl would come to depend on such patronage for their livelihood. This did not set the Duhre family apart from their peers: in the Swedish economy, based on credit and in which economic resources were scarce, social status was generally based on symbolic rather than economic capital. Credit was based on one's position in personal networks structured around, for example, kinship, gender and social status. Also, credit was not only an economic concept: to have good credit signalled that you were trusted and part of influential networks.[4] Thus, in spite of their relative poverty, the Duhres can be seen as inhabiting a social position just below newly ennobled families. Their social place is clearly manifested by their father's second marriage to Christina Wudd, the second daughter of the recently ennobled Paul Wudd. Although

---

[3] Bengt Hildebrand: "Duhre, släkt", *SBL* 11 (Stockholm, 1945), 505. I have not been able to locate the name of Duhre's mother in the records of the local parish.
[4] Klas Nyberg: "Jag existerar endast genom att äga kredit", 187.

the Duhres were not of noble standing, they possessed enough status for men of the family to marry women from the lesser nobility.[5]

There are few accounts of Anders' and Carl's lives as students. The most detailed one is a short autobiography, found in a drafted preface to a never published mathematical treatise by Anders. There, he described how he had *not* studied mathematics in Uppsala, although he considered that "I have been fully *inclined* towards *philosophical disciplines* since childhood". According to his own account, "ignited by the innate love that I carried for these *disciplines*," he started to study mathematics when he was 25 (i.e., around 1705–6).[6] At the expense of his parents, he visited Bergslagen to study mechanics and mathematics with Christopher Polhammar, as mechanically apt and inclined young men were expected to do. However, after living there for six months, Duhre ran out of funds and returned to Uppsala.[7]

The earliest document in Anders' hand was written shortly after he returned from Bergslagen. In a letter to the Bureau of Mines of 20 May 1707, he applied for the mechanical stipend in Polhammar's *Laboratorium Mechanicum*.[8] He wished to use the stipend to return to Bergslagen, where he would continue his studies with Polhammar. Again, he wrote of how he had commenced his studies of mathematics "not without my parents' great expense and inconvenience". He also explained that he had studied more "*theoria*" than he had "*praxis*", and gave two reasons for this asymmetry. First, he considered it to be easier to learn *praxis* after acquiring "a taste for *theoria*". Second, learning *praxis* was "more costly than I, a man without means, could bear."[9] In his preface from 1714, he again professed that he was too poor to make progress in "the mechanical *praxis* and the part of *physica*, which requires expenses and experiments."[10] In these letters of self-

---

[5] These marriages are discussed in Hildebrand: "Duhre, släkt".

[6] "Ifrån barndommen har iag fuller warit *inclinerad* till *philosophiska discipliner*"; "uptänd af den infödda kiärleken, som iag bar till samma *discipliner*,"; Anders Gabriel Duhre: "Förtahl. Hwilket And. Gab. Duhre är sinnad att sättia framman för sitt mathematiska wärk, som han nu hafwer under händer uti upsååt at låta och uthgå på wårt Swenska Moders mål", A 29, UUB [≈1714], unpaginated. In this manuscript, Duhre wrote that he started to study mathematics when he was 25 years old, that he was 28 when he finished studying for Polhem, that he was 31 when he started working on his mathematical work and that he was 34 at the time of writing the manuscript. Given the fact that Anders was born after Carl, who later inherited Gnista, and that Anders does not mention being an auscultator in the Bureau of Mines (which he became in 1715), it is likely that the manuscript is published in 1714 or early 1715 and that Duhre was born in late 1680 or early 1681.

[7] On Polhammar, later ennobled Christopher Polhem, as a teacher of mechanics and an ariter of mechanical aptitude, see page 100–2.

[8] Anders Gabriel Duhre to the Bureau of Mines: "Application for stipendium mechanicum", 1707-05-20, E4/120/791, Bergskollegiums arkiv, RA. Compare to my analysis of Polhammar's *Laboratorium* and his role as an arbiter of mechanical aptitude in Chapter 3, page 100–2.

[9] "icke utan mina K: Föräldrars stora omkostnad och olägenheet"; "smak af *theorien*"; "större omkostnader, än iag en medellös karl länge tåla."; ibid.

[10] "framsteg uti *mechaniska praxi* och den dehl af *Physica*, som fordrar omkostnad och *experimenter*."; Duhre: "Förtahl".

presentation, Duhre echoed Polhammar (discussed in Chapter 4): whereas theoretical talk was cheap, *praxis* and the integrated practice of the two epistemic virtues required financial support from the state. This view would also resonate in Duhre's project proposals of the 1720s.

However, Anders Gabriel Duhre did not receive the stipend in mechanics (as shown in Chapter 3, Polhammar instead favoured his own son), and he thus did not return to Polhammar in Bergslagen. He instead stayed at his family's knight's fee. There, in his own spare time and without a tutor, he studied pure mathematics from books. According to his autobiography, he "met great obstacles, and wasted much time" and he turned almost 31 before he made any notable progress. His claim that he did not receive any instruction is however inconsistent with other statements that he made. Later, in the same biography, he described his studies with the professor of astronomy Pehr Elvius (the elder), who:

> not just lent me his best mathematical books all the time, together with information on the method of how to read them quickly in a fruitful way, but who also, every time I returned the borrowed book to him, spared no effort in examining how well I understood what seemed most curious in them.[11]

At roughly the same time, Christopher Polhammar issued him a certificate of mathematical competence. This certificate very much corresponds with Anders' own descriptions of his skills, found in his stipend application in 1707 and his unpublished preface. Polhammar identified Duhre as a man of rare theoretical aptitude in "*mechanics* [and] *geometry*", and admitted having seen few Swedes with the same grasp of "ingenious *algebra*". Nevertheless, Polhammar stated that because of his lack of means, Anders unfortunately lacked knowledge of *praxis*.[12] On 26 September 1715, Anders Gabriel Duhre submitted an application to auscultate in the Bureau of Mines. In his application, he again requested the mechanical stipend. Here, instead of presenting himself in Polhammar's image – in other words, as a mechanicus who combined *theoria* and *praxis* in diligent work – he emphasised his competence in pure mathematics. "By using my humble capacity," he could "instruct those, who wish to learn the fundaments in the actual *Pure Mathematical Science*". Such knowledge, in turn, would facilitate "a greater skill in *Studium Physica* and the *mechanical praxis*", knowledge that Duhre argued depended

---

[11] "mötande swårigheter, med så stoor tijd spillan"; "icke allenast altijd låhnte mig sina besta *mathema*tiska böcker, iämbte underrättelse om *methoden* at i snarhet läsa dem med frucht, utan iämbwäl hwar gång som iag lefwererade honom samma låhntagna böcker igen, hade mödan ospard at *examinera* huru iag förstod det, som der uti syntes märckwärdigast."; ibid.

[12] "*Mechaniquen, Geometrien*"; "den sinnerijke *Algebra*"; Christopher Polhem: "Betyg för Andreas Duhre", 1712–07–30, X 241:Duhre, UUB. Interestingly, Polhem wrote this certificate the same year that he initiated his correspondence with Casten Feif and Carl XII.

partly on mathematics, partly on practice [*öfningen*].[13] Duhre, lamenting his lack of practical skill, instead offered his mathematical services to the king, Karl XII:

> Although I must confess that the lack of means, which until this day has been my steady companion, as if through force has excluded me from opportunities to train myself in *praxis*. Nevertheless, I dare not to wish anything more, than to be able to show the Royal Bureau of Mines, without sparing any effort, what I could accomplish among the sharp *ingenia*, who are determined to serve the Royal Bureau of Mines in the mining districts, or who in any other way wish to serve king and fatherland, in a capacity that require a mature mathematical knowledge.[14]

Instead of presenting himself as a man who could bridge the divide between *theoria* and *praxis*, Duhre argued that he, by virtue of his mathematical competence, could foster the young men of the Bureau into officials who embraced both these epistemic virtues. He presented the Bureau with the promise of a mathematical foundation for its work. Initially, the Bureau seems not to have been completely convinced. On 23 April 1716, it sent a letter to Polhammar discussing Duhre's application. Although they wished to promote "such a quick *ingenium*", they referred to the letter, discussed in Chapter 3, where Polhammar had suggested that Duhre would perhaps be of more use as a mathematics teacher in Karlskrona. The Bureau suggested that Polhammar should assess the possibility of giving Duhre such a position.[15] Whether or not Polhammar ever did this, Anders Duhre was accepted by the Bureau as a stipendiary in mechanics from 9 May 1717.[16]

Around this time, Anders moved to Stockholm with his brother Carl. Whereas Anders primarily studied pure mathematics, his older brother focused on mechanics. Anders authored mathematical treatises and Carl worked on mechanical inventions for the Bureau of Mines.[17] As a stipendi-

---

[13] "effter min ringa *capacitet informera* dem, som åstunda en grundelig kundskap i sielfwa *Rena Mathematiska Wetskapen*"; "större skickelighet till *Studium Physica* och *mechaniska praxin*"; Duhre: "Application to auscultate".

[14] "Ty fast än iag måste bekienna, at medellösheten, som in till denna dag, warit min stadiga föllieslagare, har lijka såsom med wåld uthstängt mig ifrån tillfällen at öfwa mig til praxi; icke desto mindre, understår iag mig intet högre önska, än med ospard flijt in för det höglofl Kongl Bergz Collegium kunna wijsa, hwad iag hos dhe qwicka ingenier skulle kunna uthrätta, som hafwa i sinnet at antingen i Bergzlagerna wara höglofl Kongl. Bergz Collegium till tienst, eller elliest i något annat sätt betiena Konungen och Fäderneslandet, i sådana beställningar, som fordra en mogen mathematisk kundskap."; ibid.

[15] "ett så snällt ingenium"; Bureau of Mines: "Letter to Christopher Polhem", 1716-04-23, I p:23 1, 101, KB.

[16] Hildebrand: "A G Duhre".

[17] The mechanical work of Carl Duhre is even less known than that of his brother. Nonetheless, there are a number of sources on his inventions from the late 1710s and early 1720s. On 21 November, Carl Duhre submitted an invention for the extraction of water from mines to the Bureau of Mines; Carl Duhre to the Bureau of Mines: "On an invention for water extraction", 1719-10-29, E4/143/177, Bergskollegiums arkiv, RA. In a letter of 23 March 1721, he again notified the Bureau of his work. And he also informed the Bureau that he had continued the

ary, Anders gave mathematical lectures in Stockholm, just as he had promised in his application. Together with his fellow auscultator, Georg Brandt, he published these lectures as the textbook *Mathesin Universalem*, discussed in Chapters 2 and 3.[18] In 1718, Anders wished to relocate to lecture at Uppsala University. In a letter of 20 January, in which he asked the Bureau to recommend him for a vacant position in astronomy in Uppsala, Anders presented his plan to transform this professorship into one that taught "a fundamental geometry, mechanics, and basic algebra together with its use, or *application*, to geometry, mechanics and useful things."[19]

When Anders did not receive the professorship, he started writing a geometrical treatise instead. In a letter to the Bureau of 29 October 1719, he asked for monetary support for his work. Here he also presented some aspects of his life in Stockholm during this time. He resided there, "in spite of the difficulties of the time" because "here live diverse lovers of useful knowledge". These "lovers" lent Duhre money to complete his treatise. At the time of writing his letter, Duhre lived in Stockholm in order to supervise the printing of his work, and he asked the Bureau to support him.[20] From 1720, Duhre held a position as "algebraist" in the fortification corps and his *Geometria* was published in 1721.[21] Nonetheless, his economic difficulties continued. In a letter of autumn 1721, Duhre solicited the Bureau for additional aid, because he was "exhausted and in great debt" after having pub-

---

work on his invention with the help of his brother Anders. In this letter, he asked the Bureau to support his work financially; Carl Duhre to the Bureau of Mines: "On an invention for water extraction, continued", 1721-03-23, E4/146/313, Bergskollegiums arkiv, RA. In a letter of 24 April 1721, Anders Gabriel discussed an invention of his and his brother Carl for the running of mills in still water; Anders Gabriel Duhre to the Bureau of Mines: "On an invention for the running of mills in still water", 1721-04-24, E4/147/141–2, Bergskollegiums arkiv, RA. Anders Gabriel Duhre also published on this invention in a printed book some years later, Anders Gabriel Duhre: *Tanckar, angående huruledes man i mangel af strömar och fall måtte med en synnerlig förmån, allestädes (hvarest stilla stående watn finnes) kunna drifwa allehanda rörliga wärck* (Stockholm, 1723).

Carl Duhre and his inventions are also discussed in two letters from the Bureau of Mines to the king in the early 1720s, where the Bureau proposes that Carl should be given a stipend due to his skill in mechanical work (Bureau of Mines to The Royal Majesty: "On Carl Duhre", 1720-01-19, 8/19, Kollegiers m fl, landshövdingars, hovrätters och konsistoriers skrivelser till Kungl Maj:t; Bureau of Mines to The Royal Majesty: "On Carl Duhre, continued", 1721-04-17, 8/19, Kollegiers m fl, landshövdingars, hovrätters och konsistoriers skrivelser till Kungl Maj:t.

[18] These lectures resulted in the publication Brandt: *Mathesin universalem.*

[19] "en *fundamental Geometrie, Mechanique,* och Grundelig *Algebra,* iemte dess bruk eller *application* till *Geometrien Mechaniquen* och nyttige saker."; Anders Gabriel Duhre to the Bureau of Mines: "On professorship in astronomy", 1718-01-20, E4/142/38, Bergskollegiums arkiv, RA, 40.

[20] "utan anseende till tydernas swårhet"; "här finnes en och annan af nyttige wetskapers elskare"; Anders Gabriel Duhre to the Bureau of Mines: "On support for a geometrical work", 1719-10-29, E4/144/512, Bergskollegiums arkiv, RA. His living in Stockholm to supervise his work hints at a wish to supervise the printing process similar to that of authors in early modern England, which Adrian Johns has discussed in Adrian Johns: *The nature of the book. Print and knowledge in the making* (Chicago, 1998), 101–3.

[21] Hildebrand: "A G Duhre"; Duhre: *Geometria.*

lished his two mathematical works "for the public good".[22] This letter is not only a sign of Duhre's poverty: publishing his mathematical works, together with his work in the Bureau and the fortification corps, had given him a wealth of symbolic capital and had made him into a trustworthy man in the eyes of important audiences. In other words: he was a man of credit. Despite his poverty, by the early 1720s, Anders Gabriel Duhre was recognised as an able mathematician and civil servant. At this time, his *Geometria* was the most comprehensive mathematical treatise in Swedish to date. Also, he taught private mathematics classes to promising students such as Anders Celsius and Samuel Klingenstierna. All in all, the position from which Duhre approached parliament in the 1720s was hardly that of a nobody.

Duhre's proposal of mathematical–œconomical redemption

In 1722, Duhre published his *Well-intentioned thoughts*, a short proposal of 28 pages. There he discussed his plans for what he called a "*Laboratorium mathematico–œconomicum*". This *Laboratorium*, which Anders wished to establish together with his brother Carl, aimed at "the cultivation of the most useful arts and sciences and their earnest use in the public as well as private œconomy".[23] Duhre invited the readers to a dialogue: he presented his work "to the judgement and consideration of reasonable people, concerning what should be considered before I venture to present such [a project] to the honourable estates of Sweden". In yet another attempt at dialogue, Duhre published another book later the same year, supposedly based on readers' comments on his earlier publication.[24]

Both Duhre's books proposed an educational institution that differed from universities. It would teach knowledge that was based on mathematical and œconomical principles. He stated that the mathematical sciences, as well as experimental physics [*experimental physiquen*], were the source of "all the other worldly sciences and arts". Such knowledge would only come to full use, however, if it were combined with "a diligent exercise in all sorts of practice and crafts". Therefore, the *Laboratorium* would teach young men mathematics together with crafts such as carpentry and turning. Men who knew how to combine these techniques would produce knowledge that was useful and beneficial to the *publicum*. Thus, Duhre's proposal was similar to

---

[22] "utmattad och i stor skuld fördiupad"; "för det allmenna besta"; Anders Gabriel Duhre to the Bureau of Mines: "Lack of means", 1721-10-18, E4/147/308, Bergskollegiums arkiv, RA.

[23] "de högnödigste wetskapers samt konsters planterande och alfwarsamma lempande, så til *Publique* som til *Privat Oeconomie*"; Duhre: *Wälmenta tanckar*, title page.

[24] "under förnuftigt Folcks opprövwande och betänckiande, öfwer det som kan wara, at wid detta påminna, innan jag bör fördrista mig sådant Swea Rikes Samtlige Högloflige Ständer Ödmiukast föredraga"; ibid. Duhre: *Förklaring*.

his earlier application to auscultate in the Bureau of Mines, as well as to his application for the professorship in Uppsala in the 1710s.[25]

Duhre's argument was hardly an original one. His proposal was based on the early modern social epistemology of mechanics, found, for example, in the coming-of-age narratives discussed in Chapter 2. The *Laboratorium* would teach boys both mathematical discernment and artisanal diligence, in order to foster them into men who submissively worked for the good of the state. Much of what Duhre suggested could also be found in the values and norms that came to be instilled in auscultators of the Bureau of Mines.[26] Furthermore, his suggestions are similar to those found in contemporary German publications, such as Tschirnhaus' widely circulated treatise on useful knowledge; the pedagogy of August Hermann Francke's orphanage in Halle, established in 1695; and Christoph Semler's plans in his *Neueröffnete Mathematische und Mechanische Real-Schule* from 1709.[27] However, Duhre's project is a very early example of an œconomical academy. In many respects, his plans for a *Laboratorium* resemble later German "model farms", such as those envisioned by Gottfried Schreber in 1763 and realised in the Friedrich Casimir Medicus' Lautern Physical-Œconomical Society in 1774.[28] Interestingly though, Duhre's *mathematico–œconomical* estate predates Schreber's blueprints by 40 years.

Issues of character and virtue took centre stage in Duhre's two publications of 1722. His presentation of himself and his students was interrelated, constituting another example of intergenerational relationships of mechanics. It was these imagined relationships – formed around virtuous knowledge – that Duhre presented to the four estates of parliament. He argued to his readers not only that he was a pious subject, but also that he could make a whole new generation of young men in his own image. Therefore, the greater part of his publications discussed who could do both mathematics and crafts, what kind of person could make useful knowledge and how Duhre's contemporaries could identify and value such a man. In his *Well-intentioned thoughts*, he summarised the problem:

> Considering that these persons [i.e., his students] should not learn crafts for the reason of becoming craftsmen, thus one must ask: after they have gained a mature knowledge in the above mentioned sciences and arts, to

[25] "alla de öfriga werldsliga wetskaper och konster"; "en flitig öfning uti allehanda *practique* och handtwärck"; Duhre: *Wälmenta tanckar*, 2.

[26] See Chapter 2 and 3.

[27] Tschirnhausen: *Gründliche Anleitung zu nützlichen Wissenschafften, absonderlich zu der Mathesi und Physica, wie sie anitzo von den Gelehrtesten abgehandelt werden. Dritte Aufflage vermehret und verbessert*; Christoph Semler: *Neueröffnete mathematische und mechanische Real-Schule* (Halle, 1709). On the educational use of practical mathematics in Francke's schools, see Whitmer: "Eclecticism and the technologies of discernment in pietist pedagogy". The parallels between Duhre's plans and contemporary German pedagogues are discussed in Wiberg: *Till skolslöjdens förhistoria* vol. 1, 30–5.

[28] Wakefield: *The disordered police state*, 114.

which ends should they be used, how should they be rewarded, and how could the usefulness of establishing such a *Laboratorium* be so great?[29]

As discussed in Chapter 2, Duhre wanted the students of his *Laboratorium* to learn the skills of craftsmen without becoming craftsmen themselves. Instead, he wished to teach "the mathematical sciences and experimental physics" to young students.[30] This was to be accomplished by an array of means, one being the translation of important mathematical works into Swedish. Without textbooks in Swedish, Duhre argued, it would "be impossible to reap their full use and utility".[31] He also argued that the education of the *Laboratorium* should be given in Swedish, and that it should be free. In this way, he hoped to attract students of modest origins who were destitute of means [*medellösa*]. Apt boys could be found in all social strata, and a broad base of recruitment would guarantee that the state used its subjects' potential as efficiently as possible. As discussed earlier, Duhre himself repeatedly presented himself as such a useful, but poor, man. When he argued that poor boys were more susceptible to bad influences than young noblemen, he thus discussed a category of boys to which he had previously presented himself as belonging. However, such students from modest backgrounds were not necessarily virtuous, because they could have been badly brought up. The *Laboratorium* thus needed to introduce poor apt boys to a virtuous way of life.[32]

For Duhre, it was "the depravations of the youth, which are either innate or have been inveterated by a bad upbringing," which were "the only and greatest obstacles to my intent."[33] Parents' neglect or inability to raise children into virtuous subjects could therefore inflict great harm on the common good. Duhre complained how "the mighty dominion of evil desires to such a degree has conquered human reason".[34] Less than virtuous actions shaped young people into bad subjects and consequently damaged the state. By neglecting the upbringing of their children, parents (and here Duhre specifically addressed women) failed in the duties of a pious subject. As a solution, Duhre recommended parents, who lacked knowledge in how to educate their child, to "read the work published in English by the ingen-

---

[29] "Såsom dessa Personer icke böra lära handtwärcken til den endan, at de skola blifwa handtwärckare; Altså frågas, sedan de uti ofwannämde wetskaper och konster, hade bekommit en mogen kundskap, hwar til de sedan skulle brukas, huru de skulle blifwa belönte, och huruledes nyttan af ett sådant *Laboratorii* uprättande kan wara så stor?"; Duhre: *Wälmenta tanckar*, 5.

[30] "de *Mathematiska* Wetskaper samt *Experimental Physiquen*"; Duhre: *Wälmenta tanckar*, 2.

[31] "omöyeligt, at deras fulla lempande och nytta kunna erhållas"; ibid., 3. Compare to the discussion of the relationship beteen Swedish and Latin in mathematical textbooks on page 51–6.

[32] Ibid.

[33] "ungdomens wanarter, som antingen äre infödda, eller igenom en ond uptuchtelse kommit at inrotas, såsom detta mit upsåts endasta och största hinder."; ibid., 25.

[34] "onda begärelsers starcka wälde så hafwa öfwerwunnit menniskliga förnuftet"; ibid., 26.

ious Englishman *John Locke*, which now, because of its great value and usefulness, is translated into our mother tongue".[35]

In *Some thoughts on education*, translated into Swedish as early as 1709, Locke presented a plan for the education of children, seen as shaped by their actions and sensual perception. He thus expressed a performative concept of self: the actions of those around them, together with their own repetitive performance, moulded children into diverse forms. Because children were formed by their parents' actions, physical punishment should only be used sparingly, and in most cases only when all other means of correction had failed.[36] In that way, children would learn how to embrace what was good, and not just to fear physical punishment. Duhre's argument, that boys who practised what was virtuous would become pious and submissive subjects, can thus be seen as an adaption of Locke into the hierarchical order of the Swedish state. According to Duhre, although "at first glance, many women might consider [what Locke argues] strange and laborious [...] they will eventually recognise how they were erroneous in their judgement."[37] Hence, the educational role that Duhre attributed women in raising a mechanicus corresponds to the fictional female characters discussed in Chapter 2 (see page 57–58). While the *via media* leading to mechanical manhood was not open to women, their hand was central in nurturing apt boys who, in time, would become virtuous and useful men.

Seeing the moral responsibility he would have for his students, Duhre did not wish to locate his *Laboratorium* in a large town full of immoral influences. In a town, "the youth may have the opportunity to associate with bad company, which is the cause of all thereof resulting vice."[38] Especially, he argued, "the children of poor people, and young students destitute of means, suffer a great risk in large towns".[39] Therefore, the *Laboratorium* should be located in the countryside. While rural areas did "not provide the opportunities to socialise with all sorts of people, which one can find in large towns", in the countryside the young could "acquire a habit of doing what is virtuous, so that they may be found, as it were, unskilled in all bad

---

[35] "läsa den myckit sinnerika Engelsmannens *Johan Lockes* der om uppå Engelska språket utgångna wärck, som nu sedermera för dess ogemena wärde och nyttighet finnes uti wårt modersmål öfwersatt"; ibid.

[36] Shapin: *A social history of truth*, 75.

[37] "fast än något der uti kan förekomma mångom af fruentimren wid första åskådande äfwen så sällsamt, och beswärligit, som de sedan måtte sådant omdömme tilskrifwa sin egen missräkning."; Duhre: *Wälmenta tanckar*, 26. John Locke: *Some thoughts concerning education* (London, 1693). The Swedish translation that Duhre refers to is John Locke and Matthias Riben (tran.): *Herr Johan Lockes tankar och anmärkningar angående ungdomens uppfostring först skrefne uti engelskan, men nu för deras serdeles wärde och nyttighet uppå swenska öfwersatte* (Stockholm, 1709).

[38] "ungdomen kan hafwa tilfälle at umgå med ondt sälskap, hwilcket är uphofwet til alla der af flytande odygder."; Duhre: *Förklaring*, 28.

[39] "fattigt folcks barn och medellösa unga emnen uti stora städer äre underkastade en myckit stor fara"; ibid. Duhre's line of reasoning is in line with a general class dimension in Swedish eighteenth-century discussions of the moral influence of towns. See Sennefelt: *Politikens hjärta*, 48.

things." By unlearning their ability to do bad actions, young men would "in time acquire the skill of being of use for the realm, and also of being agreeable in all company."[40] In Duhre's text, usefulness, virtue and social acceptance were interrelated.

As discussed in Chapter 2, Duhre presented numerous examples of immoral mathematical and mechanical practitioners. Such men were not only bad subjects, they also brought disgrace upon the mathematical sciences. It was because of these bad mathematicians that "*mathematical* knowledge is innocently subjugated to all kinds of loathsome judgements everywhere where it is not prevalent."[41] In his publications, these bad examples worked as counter-images of himself and his own project. There was always a risk, he argued, of associating mathematics with ambitious promises. Because mathematics was an "immature science in relation to its expected progress," mathematicians could "fear disapproval from all those, who do not know the great extent of its possible applications."[42] Therefore, to establish an education that fostered virtuous mathematicians, who would deliver on their promises, was also to salvage the reputation of mathematics. But were the performance of Duhre's school not to meet the expectations of his audience – expectations that Duhre consciously raised by promising great results – he could expect to meet disproval.

In his publications, Duhre's projected person, his *Laboratorium* and his students formed a coherent whole. The *Laboratorium* was presented as a virtuous place, isolated from bad influences, where Duhre – a useful and virtuous *mathematicus* – would pass on his own traits to the students. There, a whole new generation of useful and virtuous subjects would be educated in the norms and techniques by which mathematics and crafts could promote the economy of the *publicum*. After having made his argument clear, Duhre laid down his final sales pitch for the project, presenting the consequences it would have for the Swedish state:

> [A] worn out country has stumbled upon the straightest road to its redemption, after it has been advised to let well intending and wise people examine what one could possibly attain through the *application* of the two principal sciences [i.e., *mathesis* and *physica*].[43]

---

[40] "icke tillåter at hafwa den anledning til umgiänge med allehanda slag af folck, hwilcken man kan hafwa uti stora städer"; Duhre: *Förklaring*, 28–9. "til en så faststäld wana, at giöra det som dygdigt månde wara, at de måtte finnas lika såsom oskickeliga til alt wanartigt."; ibid., 29. "at med tiden för Riket blifwa gangneliga, och tillika uti alt umgiänge behageliga."; ibid.

[41] "*Mathematiska* wetskapen oskyldigt måtte wara underkastad widriga omdömmen å alla orter, hwarest den intet är giängse."; Duhre: *Förklaring*, 12.

[42] "omogenhet i anseende til dess förmodade tilwäxt"; "befruchta misshag hos alla dem, som icke weta huru wida dess *Application* sig månde sträcka."; ibid., 14.

[43] "et utmattadt Rike har råkat uppå den ginesta wägen ledsagande til dess uprättelse, sedan det kommit uppå det rådet, at låta wälmenande och klokt folck undersöka hwad man månde kunna uträtta igenom de twänne hufwud wetskapers *Application*"; ibid., 16.

By teaching the skills of craftsmen together with the "principal sciences", Duhre proposed a formal route of education for men in the Swedish state. This, in turn, would facilitate an economic redemption of the Swedish realm. Thus, what Duhre presented was not only an educational institution, but also a vision of a well-ordered state, administered along mathematical, mechanical and cameral principles. His *Laboratorium* was the model of such a state, as well as a school for the men who could fully realise the model.

The publications as performance in the parliament of 1723

The publications from 1722 were the basis for Duhre's submission to the Swedish parliament of 1723. In the spring of 1723, he presented a proposal to the four estates of parliament in which he had concretised his vision. In this proposal, he suggested that he should be awarded the royal estate of Ultuna, south of Uppsala, the revenue of which he would use to fund his *Laboratorium*. Ultuna had seen a number of tenant farmers come and go over the first decades of the eighteenth century. During the Great Northern War, Karl XII had pawned the estate to the counts Carl Gustaf and Thure Gabriel Bielke. In 1713, they had in turn transferred their contract to the Uppsala professor of theology and cathedral dean, Lars Molin. At the time of the parliamentary sessions of 1723, the county governor, Baron Magnus Cronberg, had leased the estate for only 2 years. It seems as if none of these tenants had invested much work or resources in Ultuna. By the early 1720s, it was a run-down estate, far from the visions found in Duhre's publications.[44] Its location suited Duhre, however. Not only was it close to the university in Uppsala, Ultuna happened to be located just west of Gnista, controlled by the Duhres. The royal estate and the family estate of the Duhres were separated only by the King's Meadow [*Kungsängen*]. By acquiring Ultuna, they would thus control most of the fertile fields just south of Uppsala. Also, Duhre turned the bad shape of the estate into an argument. He proposed to the parliament that, through his care, the estate could be transformed from an object that required substantial investment into a *Laboratorium* that would restore not only Ultuna itself but also the whole Swedish state.

Duhre was successful in selling this vision to his audience. In the protocols of the estates of the parliament in Stockholm, it is possible to study the reception of Duhre's proposal. On 29 March, the report of the Deputation of Cameral Matters and Œconomy [*Cammar- och œconomiedeputationen*] concerning Ultuna was read to the members of the clerical estate. This report

---

[44] Hebbe: "Anders Gabriel Duhres 'Laboratorium mathematico-oeconomicum'. Ett bidrag till Ultunas äldre historia", 576.

suggested that Ultuna, on some conditions, should be leased to Duhre "to establish a *Laboratorium mathematico–œconomicum* there" and that he should be given a privilege so that no one else should be allowed to "construct [things] after the method and design of his inventions." The estate gave its full support to the first proposal, and decided to give the second proposal some further thought.[45] The very next day, the clerical estate communicated their decision to the nobility as well as the burghers.[46]

On 4 April, the peasant estate discussed the report at some length, but decided to postpone their decision. They continued their discussion on 9 April, only to decide that they would leave the decision to the king.[47] The same day, the burghers and the nobility also discussed Duhre. The burghers noted in their protocol that the current tenant of Ultuna, Baron Cronberg, had submitted a counter-proposal concerning Ultuna to the Deputation of Cameral Matters and Œconomy. Cronberg had argued that the office of the county governor (which he inhabited) should be given Ultuna permanently. Likewise, Duhre had submitted another request for Ultuna. The deputation had again sided with Duhre, and the burghers agreed with the deputation's decision. They noted that Ultuna had never before been the residence of any governor. But more importantly: "such a Laboratorium œconomicum-mathematicum [sic], which Duhre offers to establish, is rather useful for the public". The burghers anticipated good results from the *Laboratorium*, highlighting "Duhre's known diligence and inclination to devote both his work and knowledge to the service of the publicum".[48]

The nobility was less decided. Their protocol of 9 April described Cronberg's counter-proposal in greater detail than the burghers' protocol, and the governor argued that Duhre "was more skilled in mechanics than in farming". Cronberg also argued that it would be better to give Duhre a house and a salaried position in Stockholm, where he could teach young men mathematics. In Stockholm, wrote Cronberg, "there are a number of youths who could profit thereof." While Baron Hugo Hamilton agreed that one should "support the mathematical sciences," he objected that: "one must not take someone else's property". Baron Göran Silfwerhielm answered that "If Doctor *Molin* would still have had Ultuna, no one would have talked about that [i.e., the property issues]." The elephant in the room,

---

[45] "at ther inrätta et laboratorium mathematico-oeconomicum"; "bygga effter hans inventioners methode och inrättning."; Axel Norberg: *Prästeståndets riksdagsprotokoll* vol. 6 [1723] (Stockholm, 1982), 88.

[46] Ibid., 92–4. See also Nils Staf: *Borgarståndets riksdagsprotokoll från frihetstidens början* vol. 2 [1723] (Stockholm, 1951), 100; Sten Landahl: *Sveriges ridderskaps och adels riksdagsprotokoll från och med år 1719* vol. 2 [1723] (Stockholm, 1876), 161.

[47] Sten Landahl: *Bondeståndets riksdagsprotokoll* vol. 1 [1720–7] (Stockholm, 1939), 128–9.

[48] "et sådant Laboratorium oeconomicum-mathematicum [sic], som Duhre tilbjuder sig at inrätta, är för publico ganska nyttigt"; "Duhres försporde flit och åhoga at wilja upoffra både arbete och wetskap til publici tjänst"; Staf: *Borgarståndets riksdagsprotokoll från frihetstidens början* vol. 2 [1723], 107.

which Silfwerhielm not so subtly hinted at, was the fact that Cronberg himself was a parliamentary representative of the noble estate, and this was probably why the tenantship was a more complicated matter for the nobility than for the other estates. A large number of the nobles were in favour of the report's suggestion, but others wanted to refer it back to the Deputation of Cameral Matters and Œconomy. Finally, the estate decided to convey Cronberg's counter-proposal to the other estates and to the deputation.[49]

On the morning of 11 April, the nobles had communicated Cronberg's proposal to the clerical estate, but they did not much care for it. The clerics pointed out that it had already communicated its decision on the matter, and that it would not change its position.[50] On 8 May, all four estates again discussed what was to be done about the Ultuna estate. The peasant estate stood by its old verdict: the king should decide. The burghers and the clerical estate remained positive about the deputation's proposal. Among the nobility, Cronberg was the only one who argued for his counter-proposal. The rest of the estate seemed tired of debating the issue, and after a short discussion they too decided to agree to the deputation's original proposal. Thus, after 8 May, no estate was opposed to Duhre's proposal.

Hence, the lease of the royal estate was transferred from Cronberg to Duhre. This is in itself interesting: Cronberg was a baron, a member of the noble estate of the Swedish parliament, and the county governor of Uppsala where the royal estate was situated. Obviously Duhre did not acquire the estate because of his official social rank. He was neither a member of the nobility, nor did he hold a high position in the state administration. Instead, his status was linked to his visions for an orderly state. In his publications, Duhre had woven together the *Laboratorium* with the image of himself as an ideal mechanicus and mathematicus. Judging from the protocols, he convinced the estates by successfully meeting their expectations of what constituted a mechanicus. By making himself the symbol of a dream of prosperity, he made himself worthy of the royal estate. His success was not only a result of his publications: the audiences judged these publications through their preconceptions of their author. Duhre's position as a junior official in the civil and military administration, with a reputation for mathematical skill, made it possible for him to argue convincingly for the redemptive properties of mathematics integrated with crafts. In the eyes of the estates, Duhre's mathematical and œconomical dreams seem to have held more sway than Cronberg's position or concern for Duhre's lack of agricultural experience. By beholding the future through Duhre's *Laboratorium mathematico–*

---

[49] "lagdt sig mehr uppå mecaniquen än åkerbruk"; Landahl: *Sveriges ridderskaps och adels riksdagsprotokoll från och med år 1719* vol. 2 [1723], 167. "hvarest finnes en hop ungdom, som deraf profitera kunna."; "matematiska vettenskaper böra uphielpas,"; "men man får intet taga bort en annans egendomb."; "Om Doctor *Molin* har nu haft Ultuna, så har ingen talt om det."; ibid., 168.
[50] Norberg: *Prästeståndets riksdagsprotokoll* vol. 6 [1723], 103.

œconomicum, the members of the estates could imagine a new Swedish state transformed into a well-ordered cameral machine.

The *Laboratorium* as performance in the parliament of 1726–7

In 1725, the rector magnificus and the academic senate [*konsistoriet*] of Uppsala University received notification from the Bureau of Accounts of the establishment of a *Laboratorium mathematico–œconomicum* in Uppsala.[51] Duhre had envisioned his *Laboratorium* to be hierarchical in structure – like the rest of early modern society – with craftsmen, crofters and students subordinated to its members. The members would teach the students their respective skills and be moral examples for the young men. In practice, the education that Duhre gave at his *Laboratorium* came to be structured differently from that originally foreseen in his publications. Whereas Duhre had envisioned a more or less self-sufficient school, his *Laboratorium* came to be strongly linked to Uppsala University.

In the 1720s and the 1730s, he would have assistants teach mathematics, as well as crafts such as turning and engraving, to students from the university. His initial plan was for his former private students in mathematics, Celsius and Klingenstierna, to teach at his *Laboratorium*. Because these men became occupied elsewhere, Duhre hired the mathematician Thomas Fuchs to give these lectures instead.[52] Duhre also hired the auscultator of the Bureau of Mines, Carl Meurman, who had travelled to the Netherlands to learn the art of engraving, to teach the students of the *Laboratorium* this art. (Meurman also engraved the frontispiece of Celsius' *Arithmetica*, see page 60.)[53] In the 1730s, many applicants to the Bureau of Mines had attended mathematical lectures or craft exercises linked to Duhre's *Laboratorium*. This also coincided with the increased number of applicants to the Bureau who discussed mathematics (see Chapter 3). Thus, the *Laboratorium mathematico–œconomicum* would have a noticeable impact on the Swedish mining administration, and probably on the other branches of the state administration as well[54] Using the funds of Ultuna, as well as his private savings,

[51] Kammarkollegium: "Letter to the academic senate of Uppsala University", 1725-09-04, Collegiernas bref & resolutioner för Uppsala Universitet 1623–1733, UUA, 413.
[52] Anders Gabriel Duhre: "Memorial angående Ultuna med bilagor", 1731, Frihetstidens utskottshandlingar, R 2556, RA, 83–4.
[53] Duhre: "Memorial, 1731", 97.
[54] Seventeen of the students who took classes in mathematics and crafts at the *Laboratorium* would later apply to auscultate in the Bureau. These were Olof Olofsson Kalmeter (1711–66), Johan Urlander (1712–1802), Harald Christiernin (1711–91), Johan Fredrik Ennes (1711–73), Fredrik Muncktell (1710–63), Olof Colling (?–1766), Daniel Tilas (1712–72), Ludvic von Bromell (1710–33), Gustaf Lundstedt (1710–66), Reinhold Persson Löfman (1712–68), Thomas Blixenstierna (1711–53), Carl Fredrik Geringius (1714–76), Gustaf Spalding (1709–78), Henrik Teofil

Duhre also financed a number of publications in Swedish. For example, Celsius' *Arithmetica* (1727), discussed in Chapter 2, was published on Duhre's initiative and he also funded Mört's Swedish translation from 1728 of Christian Wolff's *Entdeckung der wahren Ursache von der wunderbahren Vermehrung des Getreydes* (1719). Mört's translation of this work on agriculture was explicitly written for the Ultuna *Laboratorium*, which was noted in its title.[55]

Duhre imagined an institution where agricultural practice, crafts, mathematics and œconomy would benefit each other. However, the agricultural foundation for the *Laboratorium* would eventually generate a number of problems, after the granting of the lease of Ultuna in 1723. From the start, it was an institution marked by the friction between the grand visions of its founder and the agricultural practice through which it was funded. In his publications, Duhre downplayed this friction. He recognised the importance of agricultural skills, as well as the fact that he was no farmer himself. However, he suggested that he would initially hire a "sensible farmer", from whom he could learn the necessary farming skills. In little time and with little effort, Duhre imagined he would be able to "learn as much, as such a farmer knows". The management of the estate would then be an integrated part of the mathematical and mechanical education at Ultuna. When his mathematical–œconomical *Laboratorium* was up and running, the responsibility for the estate would be passed around, with two members of the *Laboratorium* caring for the fields one year at a time. Even though this might hamper their studies, the members would consider it a benefit. By running the estate, the young would learn the practices of private economy [*hushållningen*]. Also, by carrying out agricultural work, the members would "have their senses refreshed by a variety of useful chores, and [...] thus be a good help for their fellow brothers, who are to care fore the household."[56]

That Duhre considered farming to be an important skill for a maker of useful knowledge is unsurprising. An overwhelming part of eighteenth-century Europe was made up of rural societies structured around farming and agricultural work. In this respect, the Swedish core provinces around Lake Mälaren were typical of the rest of the continent. Because the income

---

Scheffer (1710–59), Anders Lundström (1711–93), Carl Henric Wattrang (1713–66), and Carl Anthelius (1712–?). Compare ibid., 317–31; with "Incoming correspondence from the 1730s", E4/164–92, Bergskollegiums arkiv, RA.

[55] Celsius: *Arithmetica*; Christian von Wolff and Johan Mört (tran.): *Christian Wolffs Grundeliga underrättelse om rätta orsaken til sädens vnderbara förökelse, vti hwilken tillika förklaras träns och plantors wäxt i gemen; såsom första profwet, af vndersökandet om plantors wäxt, på tyska först vtgifwen; men sedermera alla hushållare i gemen, och i synnerhet Ultuna laborator. til tienst, på swenska öfwersatt, och med hans kongl. maj:ts allernådigste privilegio vplagd.* (Stockholm, 1728).

[56] "förståndig Landtman"; "lära så mycket, som en sådan Landtman förstår."; Duhre: *Förklaring*, 40. "hafwa sina sinnens förfriskning af nyttiga sysslors ombyte, och [...] således wara sina medbröder, som skola skiöta hushållningen, til en god hielp."; ibid., 41.

from the Ultuna estate was the economic foundation of Duhre's *Laboratorium*, and agricultural reform was one of its goals, farming skills became central to Duhre's ability to function as a director of his *Laboratorium*. In turn, these skills became part of his performance as a useful subject. In order to be deemed competent in running Ultuna, Duhre had to perform as a competent master of the household [*husbonde*] for the parliament and state administration in Stockholm. When they offered him the lease, these audiences also raised their expectations of him. Thus the initial success of his publications did not only mean that he gained resources for his project, but also that he was now required to add a new repertoire to his performance as a useful subject. Duhre found himself relocated: from the well-defined hierarchical framework of a junior official in the Bureau of Mines to the position of a master of a royal estate with responsibilities not only for his *Laboratorium* but also for servants and crofters.

At first, Duhre's reputation as a virtuous man was a resource at Ultuna and his early years seem to have passed relatively smoothly. He developed the Ultuna estate; he constructed new buildings, such as a new smithy; and he managed to attain the lease of the King's Meadow, in between Ultuna and the Duhre family estate of Gnista. At the parliamentary session in Stockholm 1726–7, it was clear that he had consolidated his position as a well-respected and competent man. In 1726, he submitted a new proposal to the estates, comprising 31 pages in small print and entitled "And. Gab. Duhre's Humble proposal, directed at the highly praised estates at the parliament of 1726, concerning some of the means by which he and his brother Carl Duhre, as through a small-scale model, aim to show the possibility of, without a cost to the public, establishing the Ultuna Society, along with a report on what he has accomplished since [taking over] the royal estate of Ultuna". This publication was longer than those of 1722; in detail it enumerated what Duhre had accomplished so far and what he wished to accomplish at Ultuna in the future.[57] In his proposal, Duhre argued that the short-term lease should be made permanent. On 16 November 1726, he even presented this proposal to the clerical estate in person.[58]

The estates were generally in favour of Duhre's new proposal; their reaction was clearly based on his reputation as a man of action who could be trusted to act in the interest of the state. The parliamentary protocols from this period are filled with praise of his project. In July 1727, a new

---

[57] Anders Gabriel Duhre: *And. Gab. Duhres Ödmiukaste memoriale ständt til Swea rikes samtel. högloflige ständer wid 1726 åhrs riks-dag* (Stockholm, 1726). The full original title, quotes above, reads "And. Gab. Duhres Ödmiukaste memoriale ständt til Swea rikes samtel. högloflige ständer wid 1726 åhrs riks-dag, om en deel af de utwägar hwar med han och desz broder Carl Duhre lika såsom med en modell uti litet emna wisa giörligheten. Af, at utan det almännas betungande uprätta Ultuna societet, jemte berättelse om det, som den förra ifrån sit avträde til Ultuna kongz ladugård kan hafwa uträttat".

[58] Axel Norberg: *Prästeståndets riksdagsprotokoll* vol. 7 [1726–31] (Stockholm, 1975), 106.

proposal from the Deputation of Cameral Matters and Œconomy suggested that he should be given a lifetime lease of Ultuna. The peasant estate were generally positive about this suggestion, mainly because of their view of Duhre's character, "who in this institution does not have any motive of personal gain but, oppositely, has the honourable intent to use all his time, reason and earthly possessions to the service and use of the publicum".[59] In the noble estate, Tomas Fehman described Duhre as "an example without example, that he not only makes himself a slave of the public, but also gives away what he might win." On the other hand, Jacob Grundelstierna gave a word of warning: "To this date, he has not performed, but has only hawked hay and other things from the estate."[60] The clergy estate were also positive in its reaction, but wished for Duhre's project to be better evaluated.[61] Similar praise of Duhre can, moreover, be found in the preface to the Swedish translation of Johan Friedrich Weidler's *Institutionibus mathematicae* from the same year. The translator Mört, who later the same year was accepted as an auscultator in the Bureau of Mines, described "*Anders Gabriel Duhre's* praiseworthy intent, to serve the public good," his "beautiful works with useful proposals for the development of the manufactories," as well as how he "wanted to give any young man who is so inclined, the opportunity carefully to learn all sorts of *mathematical* things, together with the proper exercise of *practice*."[62]

In 1727, Duhre was thus perceived as an incarnation of the good mechanicus. He embodied positive values such as public spiritedness, diligence and mathematical creativity; he was considered the herald of an affluent future; and he was seen as a guarantor of a well-ordered state. Although he initially managed to live up to these high expectations, nevertheless he would soon be criticised for not meeting his audiences' rising demands. Between the end of the parliament of 1727, and the next session in 1731, a number of events would unfold that would radically change his authority to weave mechanical and mathematical dreams.

---

[59] "som vid detta värk icke har någon afsicht på sin enskylte förmon utan tvert om et berömvärdt upsåt at använda all sin tid, förstånd och timmeliga välfärd til publici tienst och nytta"; Landahl: *Bondeståndets riksdagsprotokoll* vol. 1 [1720–7], 677.

[60] "et exempel utan exempel, at han icke allenast giör sig til en träl för publico, utan ock skiäncker bårt det han kan förvärfva."; "Härtil dags har han intet præsterat, utan bara månglat ut höö och annat som fallit vid gården."; Sten Landahl: *Sveriges ridderskaps och adels riksdagsprotokoll från och med år 1719* vol. 5 [1726–7] (Stockholm, 1879), 398.

[61] Norberg: *Prästeståndets riksdagsprotokoll* vol. 7 [1726–31], 393.

[62] "*Anders Gabriel Duhres* berömliga upsåt, at tiena det allmänna bästa,"; "wackra Skrifter och nyttiga förslag til *Manufacturernas* uphielpande,"; "will gifwa hwar och en af Ungdomen som sielf har lust, tilfälle at få grundeligen underrätta sig i hwarjehanda *Mathemati*ska saker, och der jämte behörigen öfwa sig uti *practiqven*."; Weidler and Mört (tran.): *En klar och tydelig genstig*, preface. In his application to the bureau, Mört discussed his skill in mathematics, experimental physics and chemistry. See Johan Mört: "Application to auscultate", 1725-01-22, E4/159/912, Bergskollegiums arkiv, RA.

Dirty stories of the bad mechanicus

From 1727, things went downhill fast. In the parliament of 1731, Duhre's character, and the good intentions of his project, began to be questioned. Soon he would be more similar to the bad mechanicus that he, in his publications of 1722, had put forth as his opposite, than to the man who would save the Swedish state. In order to understand his transformation from a mathematical and œconomical saviour of the realm to an impostor selling false mathematical dreams, we must focus in on the local community around the Ultuna estate. The regional archives in Uppsala, as well as the national archive in Stockholm, contain a number of documents from the years 1727–31, which show a Duhre who had diverged radically from the man in the written sources discussed earlier.

On 19 September 1728, the hundred court [*häradsrätt*] of Ulleråker, with jurisdiction over Ultuna, was in session. Among its cases was one that involved Duhre. According to the county sheriff [*länsman*] Petter Hartman, during Michaelmas of 1727 – just after the parliamentary session in Stockholm and while Mört was writing his celebration of the *Laboratorium* – Duhre had repeatedly fornicated with Anika Jeansdotter[63], a maid at Ultuna. These acts had not only been committed "in [Duhre's] bedroom there at Ultuna, but also in the cowshed, so that he had never left her alone or had given her any peace". On 13 July 1728, Anika gave birth to a child, whom she argued was Duhre's bastard.[64] This court case, of fornication in a cowshed, might seem far from the epistemological issues of Duhre's mathematical and mechanical work. However, because early modern mechanics was performed in relation to a social epistemology where virtue and usefulness were interwoven, this story is of more relevance to Duhre's visions than one might first believe.

In the seventeenth- and eighteenth-centuries, natural philosophy was performed in households.[65] There, women and children partook in knowledge making, conducted in homes controlled by male natural philosophers. But the household also played an important part for an unmarried

---

[63] Unlike the men mentioned in the protocol, Anika Jeansdotter's name was not written in Antiqua to make it stand out. Instead, it was written using the same German handwriting used in the rest of the text. Most likely, this reflected her low status, as a woman and a commoner. Unfortunately, this also means that her surname is unclearly written and hard to read. "Jeansdotter" is my interpretation of it.

[64] "icke allenast inne uti hans Cammare der på Ulltuna, utan och i Fähuset, så att hon alldrig för honom fått wara i fred och oförsökt"; "Dombok från Ulleråker ting (1711–29)", A1:2, Ulleråkers häradsrätt, ULA.

[65] See for example Gadi Algazi: "Scholars in households. Refiguring the learned habitus, 1480–550", *Science in context* 16:1–2 (2003), 9–42; Frances Harris: "Living in the neighbourhood of science. Mary Evelyn, Margaret Cavendish and the Greshamites", in Hunter and Hutton (eds): *Women, science and medicine 1500–1700*, 198–217; Frances Willmoth and Rob Iliffe: "Astronomy and the Domestic Sphere. Margaret Flamsteed and Caroline Herschel as Assistant-Astronomers", in Hunter and Hutton (eds): *Women, science and medicine*, 235–65.

male mathematician such as Duhre. Duhre had to perform as a good master of the household, in order to be recognised as a credible knowledge maker and educator. In this context, being unmarried was a problem.[66] In his *Thoughts on mechanics* (1740), Polhem elaborated on the mechanical practitioner as a selfless subject.[67] In this text, published in the mid 1700s, Polhem had adapted to the new constitutional regime. Like many others before him, he explicitly linked *œconomy* to mechanics through a marital metaphor, interweaving the *publicum* (i.e., state) and the private households of the realm. Public and private economy (*oeconomia* and *hushållning*) depended on each other "like a husband and his wife". The *publicum* could be considered the man of the house: it cared for the private households of the realm from a patriarchal position. But it also depended on the orderly and virtuous conduct of the subordinated private households. Polhem considered mechanics to be the servant of this patriarch, as were geometry, chemistry, botanical anatomy, and other useful sciences and arts.[68] Thus, Polhem considered the value and authority of mechanics to come from its submission to the ultimate power. He explained the epistemic purpose of mechanics through a marital and patriarchal metaphor: mechanics should bring *theoria* and *praxis* into matrimony, for the benefit of their master. Being wed, the two would be able to produce legitimate children instead of bastards; in other words, they would produce useful men and machines built according to a plan instead of just haphazardly.[69] In Polhem's *Thoughts on mechanics*, the social order of good government – both of the state and the private households – was envisioned as a social–epistemic order where the relationship between *theoria* and *praxis* was similar to that of a husband and a wife in a private household.[70]

The house and its master were central in the agricultural economy and the patriarchal ideology of eighteenth-century northern Europe.[71] The house and the *Hausvater* or *Husbonde* (i.e., the master of the household, or husband)

---

[66] Lyndon Roper's account of the gender relationships in the Holy Roman Empire in this period is also applicable to Sweden: "The real man was a household head, a little patriarch ruling over wife, children, servants, journeymen and apprentices"; Roper: *Oedipus and the Devil*, 46.

[67] On the patriarchal order of early modern Sweden, see also Chaper 2, page 42–44.

[68] "som mannen med sin hustru."; Polhem: "Tankar om mekaniken", 188. Polhem's analogy between state and marriage follows a long tradition using the same language. See Scott: "Gender", 1070–5. For a discussion of how the early modern state was conceptualised through family metaphors that at the same time "influenced the state model of political power in the making", see Sarah Hanley: "Engendering the state. Family formation and state building in early modern France", *French historical studies* 16:1 (1989), 26.

[69] Polhem: "Tankar om mekaniken", 190.

[70] Similar relationships between theoria, praxis, husbands and wives can also be found in other texts by Polhem/Polhammar. See for example Polhem: "Samtahl emällan fröken Theoria och byggmästar Practicus"; Polhem: "Samtahl och discurs emellan theoria och praxis om mechaniska och physicalska saker huar igenom ungdomen, som der till har lust, kan lära något".

[71] Karin Hassan Jansson: *Kvinnofrid. Synen på våldtäkt och konstruktionen av kön i Sverige 1600–1800* (Uppsala, 2002), 80.

were the chief representations of domestic groupings in the Germanic and Scandinavian world, until the late eighteenth century when the concept of the *family* took hold. The "house" differed from the family in that it included servants and focused not only on the relationships of husband–wife–children.[72] Andreas Marklund has shown how the house can be characterised as consisting of three power relations: between the husband–wife, the parent–child and the master–servant. It was through these relationships that Polhem described the role of the mechanical practitioner in the state. To use a concept minted by Sarah Hanley, Polhem presented mechanics as a part of the Family–State compact, in which concepts of public power, the private household and mechanics legitimised each other. Similarly, in Duhre's case, how the power relations of Ultuna's household were perceived mattered to how the *Laboratorium* performed as a model of an ideal political order.

Duhre did not marry until 1731, when he was 51 years old. As a junior civil servant in Stockholm there was no possibility, or necessity, for him to have a wife.[73] Therefore, while the Ultuna estate household was made up of the patriarchal relationships between Duhre and his subjugated servants and young students, it lacked the relationship between a husband and a wife. Because of the absence of such a third power relation, the other two constitutive power relations of Duhre's household were ambiguous. This ambiguity also gave further meaning to the fornication court case. In eighteenth-century Sweden, as in other parts of Europe, infidelity and extramarital sex were considered threats to social order. They could bring about uncertainty concerning fatherhood, as well as unwanted economic liability for households or whole communities.[74] At a time when there was an ideological parallel between the state and the household, domestic disorder was a threat to the social order.[75] In the case of Ultuna, which Duhre presented as a model for a whole cameral order of society, such acts of vice undermined the very socio-epistemological foundation of the *Laboratorium*. If Duhre's household literally produced bastards, this in turn reflected badly on his perceived ability to perform the epistemic virtues of *theoria* and *praxis*, in order to produce virtuous men, useful machines, and an orderly and prosperous state.

Duhre had submitted a written response to the hundred court, where he disputed Jeansdotter's claims. He "denied having had any improper relations or carnal intercourse with this woman". Instead, the story "had been all made up, out of anger against him". He also argued that he had not been at Ultuna at the time of conception, "that is, when the pigs there at Ultuna were slaughtered". From Michaelmas until 5 November, he had been in

---

[72] Marklund: *In the shadows of his house*, 11. See also Dieter Schwab: "Familie", *Geschichtliche Grundbegriffe. Historishes Lexicon zur politischen-sozialen Sprache in Deutschland* (Stuttgart, 1975).
[73] Marklund: *In the shadows of his house*, 64. Hildebrand: "A G Duhre".
[74] Karin Hassan Jansson: "Ära och oro. Sexuella närmanden och föräktenskapliga relationer i 1700-talets Sverige", *Scandia* 75:1 (2009), 31.
[75] Jansson: *Kvinnofrid*, 81.

Stockholm "to deal with some personal affairs". Also, he contended, "during her service, this woman had often been in the farm-hands' quarters and giggled and enjoyed herself with the men there". Duhre argued that probably one of the servants of the estate was the child's father. He proposed that the maid accused him, instead of the real father, because he would be able to support her financially. Alternatively, he suggested, the father might be a married man, whom she did not dare to reveal out of fear of a harsher punishment.[76] Thus, Duhre paradoxically based his defence on his problematic position as an unmarried master of an estate. Being the tenant of a royal estate made him a plausible target for women who wanted support for their bastard children and his status as unmarried further facilitated such claims, because of the laxer penalties for fornication compared with adultery.

Duhre called four witnesses to his defence: Christopher Hörner, a clockmaker from Uppsala; his wife Margareta Mejer; Gördid Matzdotter, the wife of the carpenter at Ultuna; and the maid Kierstin Matzdotter. In October 1727, Hörner had constructed a clockmaker's workshop at Ultuna and during that period his wife had been a householder on the estate. Hörner pointed out that at this time, "Duhre had mostly been away," but that he "however, sometimes briefly had come home for a day or two". Nevertheless, none of these witnesses could testify to Duhre's absence at the beginning of November. Instead, the maid Kerstin told the court how "Duhre one evening, at two times had called Anika into his chambers to blow up the fire, and that she had remained in there for a while". Kierstin did not know "if something had happened between them" at that time. Apart from Duhre's witnesses, the court called two midwives, Maria Månsel and Kiersten Ersdotter, who had delivered the bastard child. The two women described Jeansdotter's childbirth as difficult. Throughout the delivery they had repeatedly asked her who the father was, and she had consistently maintained that it was Duhre.[77]

Things did not go in Duhre's favour. The case continued at the next court session on 18 January 1729. Here, Hartman charged Duhre to free himself by taking a personal oath to his innocence [personlig edgång], which was a common procedure for men involved in fornication cases. Duhre declined to take the oath, and thus he and Anika were found to be guilty. To the modern observer, Duhre's choice might seem illogical. If his innocence was just some words away, and he had spent the previous session arguing

---

[76] "nekade sig haft med denne qwinsperson något olofl:t omgiänge och kiötzlig beblandelse"; "af arghet wara på honom opdichtat"; "nembl:n då Swinen der på Ulltuna Slachtades"; "för några sine angielägenheter"; "att som denne Qwinsperson mycket trägit under sin tjänstetid wistatz i drängestugan och der med drängarne gas[s]at och flesat"; "Dombok från Ulleråker ting (1711–29)".

[77] "H:r Duhre merendels warit borto; dock ibland då och då som hastigast på en dag el:r par kommit hem"; "Hr Duhre en Afton 2:ne Ggr ropat Anika in at blåsa på Eelden, då hon någon stund drögt der inne"; "om då något skiedt dem emellan, wet hon intet"; ibid.

for his innocence, how come he now chose to remain silent? Karin Hassan Jansson has discussed a rape case from 1781, involving a farmhand named Carl, which casts light on Duhre's silence. Hassan Jansson shows how the court had a strong suspicion of guilt, but found the evidence inconclusive. Therefore, the farmhand should ideally have been forced to give an oath to his innocence. However, because of his bad reputation, the court feared that he might perjure himself. Therefore, it did not let him swear to his innocence. Instead it freed him on the grounds of inconclusive evidence.[78] The case of the farmhand highlights how, in Swedish eighteenth-century provincial courts, it could be better to free a man of bad reputation than to risk him telling lies under oath and consequently undermining the reputation of the court as a place of truth. For Duhre, the case was reversed: he was a professional truth-teller and his reputation was built around him as a virtuous man who provided useful knowledge to the *publicum*. In such a position, to accept a ruling that he opposed was better than to risk arousing suspicion of having lied under oath. Thus, on 18 January, Duhre and Anika were found guilty and charged with fines: Duhre with 10 silver dalers and Anika with twice the amount (i.e., 20 silver dalers). Duhre was judged to be the father, who should consequently pay Anika support. Also, one Sunday each, the two were to stand on the kneeler [*pliktpall*] (i.e., on the two or three steps in church where people found guilty of fornication were to stand or kneel to be pilloried and shamed during mass).[79]

This case of fornication underlines Duhre's new responsibilities when he became the master of Ultuna. It no longer sufficed for him to publish books promising the immense rewards of the *Laboratorium*: he needed to display that he could handle the complexities of running a royal estate. It also shows that the social dynamics of the Ultuna estate were not disconnected from Duhre's performance as a mechanicus. In a culture where virtue and usefulness were intimately connected, and where Duhre gained a royal estate on the basis of his good character, a mechanicus on the kneeler was a bad mechanicus.

Violent stories of the bad mechanicus

Duhre's failures as a master of an estate did not stop with fornication: he was also charged with abusing his servants. In early modern Sweden, the master of an estate had the right to administer physical punishment to his subordinates, whether they were his immediate family or his servants. But punishments needed to be perceived as just, and the law separated lawful

---

[78] Jansson: "Ära och oro", 34.
[79] "Dombok från Ulleråker ting (1711–29)"; "Plikt, 7, Plikt-pall", *SAOB* <http://g3.spraakdata.gu.se/saob/> [accessed 4 April 2015].

punishment [*aga*] from physical abuse [*misshandel*]. To cross this line was to stop being a master and start being a tyrant of the house [*hustyrann*]. As Andreas Marklund points out, in its capacity of underscoring husbandly authority "the House was not a physical entity: it was a field of responsibility, a political mandate."[80] In many ways the role of the good master and the virtuous civil servant thus intersected: like bureau officials, masters of estates were responsible for a small part of the Swedish kingdom.[81] The master of the house was supposed to police his small domain, and to maintain its order. He should not only control his own impulses, but also moderate his use of violence in order to control subordinates. Consequently, his position carried an authority based on violence exercised reasonably. In practice, however, there was a thin line between "disciplinary flogging and brutish excesses."[82] As discussed earlier, in early modern Europe the orderly house corresponded to an orderly society, and a master of a house who failed to control his dominion – either because of too little or too much violence – was also a danger to the social order. This was even clearer in the case of Duhre, who was a tenant of a royal estate, because he literally managed royal property. By parliamentary decree, his mandate over the Ultuna household should be policed by all the administrative bureaus. To these audiences, he was expected on the one hand to show that he did not intend to put his own interests before those of the *publicum*, while on the other hand he had to manifest his autonomy and his ability to perform the duties of a master. These demands arose from elevated expectations following his successful publications of 1722.

Attached to a proposal, handwritten by Duhre for the parliament of 1731, are copies of documents from another session of the hundred court in Ulleråker. Although these documents were selected and copied by Duhre and thus might be incomplete, his choice to include them in his proposal can also be seen as part of a performance made in relation to audiences in Stockholm. As such, it shows that the events at the hundred court had implications for what image his *Laboratorium* projected to the estates. The attachments represented a wide range of perspectives, many which are very critical of Duhre. These sources can thus be interpreted as several layers of

---

[80] Marklund: *In the shadows of his house*, 67.

[81] Ibid., 71–2. Compare to my discussion of the patriarchal order of early modern Sweden, page 42–44.

[82] Ibid., 72. The control of impulses and use of violence has been analysed to a large degree by historians of masculinity. On violence and masculinity in Sweden, see David Tjeder: "Konsten att blifva herre öfver hvarje lidelse. Den ständigt hotade manligheten", in Berggren (ed): *Manligt och omanligt*, 177–96. In an English early modern context, the complex and contradictive nature of impulses, violence and authority, expected from heterogenous masculinities, has been discussed by Philip Carter: "James Boswell's manliness", in Tim Hitchcock and Michèle Cohen (eds): English masculinities 1660–1800 (London & New York, 1999), 111–130. For a discussion on control of impulses and self-fashioning in English early modern science, see Golinski: "The care of the self and the masculine birth of science".

performances made by a variety of actors – locally in Ultuna, in the town of Uppsala and in parliament in Stockholm – which intersected in several complex ways. These overlapping performances carried a number of implications for Duhre's ability to present himself as a mechanicus.

It was a clerk by the name of Eric Hanssen who brought assault charges against Duhre to the attention of the hundred court in Ulleråker. In a letter to the court, faithfully copied in Duhre's proposal to the parliament, Hanssen claimed to be the spokesman for a large number of Duhre's crofters. He had compiled a list of complaints against Duhre, which the crofters were claimed to have made, and had sent it to the new county governor in Uppsala, Johan Brauner.[83] Duhre heard of these complaints from the governor on 2 August 1730.[84] In his letter, Hanssen argued that Duhre had unlawfully seized his crofters' farm animals, which were grazing in the King's Meadow. When the crofters had asked to have their animals returned, Duhre supposedly had beaten them without due cause. The crofters begged that the governor command Duhre to return the animals and that "Duhre may henceforth be bound not to give us blows and lashes, so mercilessly, at his own pleasure and after his own hard heart". Instead of having the right to inflict physical punishment, the crofters argued that the court should "let the criminal penance according to the law, and not through [Duhre's] blows".[85]

In his written response, Duhre described the accusations as preposterous and completely unwarranted. The crofters had been "deluded into having an unfounded suspicion" of him and they had "let themselves be persuaded to present a punishable insubordination to me, their rightful master". Also, the crofters' written complaints were "an unfounded petition in which I, without the slightest reason, was attacked with many insulting accusations and statements".[86] Duhre was convinced that the crofters would see the error of their ways and that they would beg for forgiveness. Instead, on 14 August, he was called to the governor for a meeting with Hanssen. At this meeting, Duhre was confronted with a second written complaint, which, he stated, "was even more offensive than the previous one".[87]

---

[83] In 1729, Johan Brauner replaced the former county governor Cronberg. Thus, like his predecessor he seems to have had a personal interest in the Ultuna estate.

[84] Duhre: "Memorial, 1731", 237–40. The complaints by Duhre's crofters can be seen in the lights of other social conflicts between lords and servants in early modern Sweden. See Ulla Koskinen: "'Benevolent lord' and 'willing servant'. Argumentation with social ideals in late-sixteenth-century letters", in Karonen (ed): *Hopes and fears*, 55–76.

[85] "Duhre må blifwa förbuden, at ey understå sig med hugg och slag så obarmhertelig oss hädan efter *tractera* efter sit egit godtycko och hårda sinne"; "låta den brotslige plikta efter lag och ey med slag"; Duhre: "Memorial, 1731", 238.

[86] "förledna til en ogrundad misstancka"; ibid., 158. "låto sig öfwertalas, til at först emot mig, som war deras rätta husbonde, förklara en straffwärdig olydno"; ibid., 159. "en ogrundad klagoskrift, hwaruti iag utan någon den ringesta orsak blef angripen med många skimfeliga beskyllningar och utlåtelser"; ibid., 160.

[87] "wara mera anstötelig än den förra"; Duhre: "Memorial, 1731", 162.

In their second complaint, the crofters accused Duhre of having called them "wild beasts and senseless brutes" and that "he seeks all sorts of pretexts of justice to profit and enrich himself from what is rightfully ours". He had denied them "the free use of our pastures, for grazing as well as hay-making, and instead let the crofters of his tax farms have their animals at the King's Meadow", actions that were "both unreasonable and indisputable". Furthermore, the complaint speculated that Duhre had been "in some confused and unfortunate state", and that he had therefore "been inclined to such unchristian actions".[88] The crofters questioned the core components of Duhre's character. Not only did being accused of being "out of his senses" undermine his self-presentation as a man of reason, it was also suggested that he put his personal gain before the well-being of the estate; that he treated his subordinates badly, verbally as well as physically; and that he was "confused" and "unchristian". Unsurprisingly, Duhre considered these accusations "awful remarks and insinuations concerning [his] office, morals and reputation".[89] The crofters had presented him as a man who could control neither his servants nor his own sensuous impulses. Thus, their insubordination jeopardised the virtuous performance of Duhre and his *Laboratorium*. The boundary between just and unjust violence did not only matter for Duhre as a master, but also for his role as an educator. As seen earlier, when Duhre argued in parliament for the importance of good morals among the youth, he recommended Swedish parents to read a recent translation of John Locke, who argued that parents should restrain from corporal punishment in most areas of a child's education.[90] If Duhre lacked the ability to distinguish between abuse and just punishment for his servants, could he be trusted with the education of young men and to be sincere in his educational programme? How he was perceived as performing in master–servant relationships corresponded to how he was trusted with giving his students a good education.

The fact that it was his servants who accused him added to Duhre's burden. The projected image of Duhre and his *Laboratorium* consisted of the collective performances of the Ultuna household. If these performances were harmonious, audiences in Stockholm and Uppsala would perceive it as a useful and virtuous place. The performances of the estate could be seen as those of "a set of individuals whose intimate cooperation is required if a given projected definition of the situation is to be maintained."[91] The serv-

---

[88] "wildiur och oförnuftiga bestar"; ibid., 246. "han söker under allehanda skien af rät kunna *profitera* och rikta sig af det os med rätta tilkommer"; ibid., 242. "at förneka os wåra enghagars frija nyttiande, så wäl uti gräsbetande som höets bergande, och i det stället låta sina torpare på des Skathemman hafwa sina Creatur i Kongs engen, är både dels oskiäligt dels och *odisputerligit*"; "i något förwirrat och beklagl. tilstånd"; "et så ochristel:t *procederande*"; ibid., 241.

[89] "hämska *expressioner* och *insimulationer* [sic] rörande mit embete, seder och *existimation*"; Duhre: "Memorial, 1731", 162.

[90] Duhre: *Wälmenta tanckar*, 25.

[91] Goffman: *The presentation of self in everyday life*, 108.

ants' signatures made such a harmonious performance impossible; instead, they projected a disorderly house, where the master could not control his subordinates. What was at stake in the court case was thus ultimately the reputation of the *Laboratorium* and Duhre's character. Having his crofters turned into adversaries broke the collective performance and undermined the presentation of self that Duhre projected to his audience in Stockholm. Thus, simply proving the crofters wrong was not sufficient. If he wished to re-establish the team performance, he had to prove that the accusation had never been made.

Duhre recognised that the acting out of his servants could have grave repercussions. In a letter to the court, he described how "lovers of the Ultuna institution" wished him to report what had happened to the "due instances" (i.e., the bureaus in Stockholm). Also, he warned about the "dire effects the insubordination of the crofters, subjugated to me, would have for all masters of households, and the likes, if it is not corrected with a proper punishment in order to scare and warn other bad servants".[92] This was a matter of "defence of public security" and on 24 August he submitted a request to the county governor for a proper judge to examine the case.[93] Hence, Duhre not only argued that the crofters were wrong in what they claimed – he also argued that their actions were a threat to the social order. The court case was not just a quarrel within the Ultuna estate: it was an event that could possibly disrupt the whole realm. Ultimately, it broke the ideal patriarchal order of the early modern state, in which all subjects piously performed according to their place in relationships of subjugation and superiority. Therefore, Duhre left Uppsala for Stockholm, where he would "turn to his royal majesty in the deepest humility".[94] To the king, Duhre recapitulated what he had written to the county governor, and asked that the Court of Appeal [*hovrätten*] should appoint a group, which Duhre would fund, that would examine the claims of the crofters.[95]

Duhre's request led to extraordinary sessions of the hundred court on 1 and 5 December the same year, to which he, Hansen and the crofters were called. There, the crofters surprisingly denied ever even having heard of their own accusations. In his proposal to the parliament of 1731, Duhre described how "when the crofters, whom I had subpoenaed, [...] were to prove [their] unfounded words of abuse [...], the correct context of the matter was finally revealed". Suddenly, it was evident that "these simpletons knew not a word of what was written in their so called complaints, which to

---

[92] "Ultuna wärckets elskare"; "wederbörlig ort"; Duhre: "Memorial, 1731", 162. "hwad skadelig påfölgd denna mina underhafwande Torpares olydno månde hafwa för alla husbönder och för dem som stå uti deras ställen, om den med behörig näpst ey skulle blifwa rättad, andra wanartige tienare til sky, skräck och warnagel"; ibid., 162–3.
[93] "almenna säkerhetens förswar"; Duhre: "Memorial, 1731", 163.
[94] "i diupaste underdånighet wända mig til hans Kongl. Maij:t"; ibid., 164.
[95] Ibid., 250.

a degree concerned the duty of my office, my honour and my reputation". While the crofters "were not without spite" towards him, they had allowed this resentment to be "put in their heads".[96] Consequently, Duhre accused Hanssen of having compelled the simple crofters to submit complaints they did not understand. He argued that Hanssen was personally responsible for what had previously been considered the crofters' list of complaints. Duhre did not only want the court to convict Hanssen for the actual words of abuse, but also to establish that Hanssen alone had made the accusations.[97] By freeing the crofters from guilt, and by undoing Hanssen's status as their spokesman, Duhre could re-establish the order of Ultuna and the virtue of his person.

Hanssen wished to defend himself against the accusations in the coming regular winter session of 21 January 1731. However, he failed to make an appearance there, and thus the court ultimately ruled in Duhre's favour. The court convicted Hanssen on several accounts. Among other things, first, he had inflamed the crofters. Second, he had "called Mr Duhre unchristian and the worst tyrant against his crofters, as if he were bothering them beyond their ability". The court pointed out that "all the crofters give high praise to Duhre, and that each man, who knows him, will assume his innocence, that he does not want to act unjust against any lesser man". Third, Hanssen had described Duhre as "out of his senses". Fourth, he had misled the crofters to claim privileges that "are proper for no subject of the realm to demand". Also, he had "taunted and criticised the royal majesty as well as arrangements and decrees made by the estates at a public parliament, by which Duhre has been given the lease of Ultuna for life, by writing that Duhre attained this benefit before other [better] qualified". Finally, which underlines how this case was connected to Duhre's *Laboratorium*, the judge ruled that Hanssen had "written sneerfully and scoffingly of the œconomical attempts that Duhre projected for the betterment of the œconomy of the realm".[98]

---

[96] "det kom der til at dessa af mig lagligen instämda Torpare [...] skulle bewisa de ogrundade tilmälen [...], yppades först sakens rätta samanhang"; "at dessa enfaldiga stackare icke wiste et ord af det som stod i bem:te deras så kallade beswärspunchtar, och i någon måtto rörde min embets plicht, heder och *existimation*"; "intet woro fria ifrån den ilwillia"; "sig låtit intala"; ibid., 173.

[97] Ibid., 257–8.

[98] "Det Cammarskrifwaren H:r Eric Hanssen uphissat de enfaldige torparne och brukt dem at underskrifwa sådane saker, som de aldrig honom ombedit at angifwa, ia intet til en del hört sig dem föreläses."; "Utropat H:r Duhre såsom ochristl:ig och den wärsta tyran emot sine torpare, som skulle han plåga dem öfwer deras förmåga"; "samtl:e torparne gifwa H:r Duhre gådt loford, och hwar man, som honom kiänner, skal anta hans oskyldighet, at han intet wil giöra någon orätt emot mindre någon menniskia"; "som wore han ochriste och ifrån sina sinnen"; "ingen undersåtare i riket skal anstå at begiära"; ibid., 259. "tadlat och *carperat* Kongl. May:t och Riksens Ständers på en almän Riksdag allernådigst giode anstalt och förordning, hwarigenom Ultuna är uplåtit H:r Duhre til arrende uti des lifstid, i det han skrifwer, at H:r Duhre kommit til en sådan

What the court's ruling makes clear is that we cannot separate the agri-cultural economic foundation of Ultuna from Duhre's presentations of himself, and the *Laboratorium*, to parliament in Stockholm. Not only was the agricultural practice planned to be a part of Duhre's education of his makers of useful science. The actions of the crofters of his estate were a part of Duhre's performance as a mechanicus. Obviously, the estate was the economic foundation of his *Laboratorium* and the controversies undermined this foundation. If Duhre failed as the master of Ultuna, he would not be able to finance his educational institution, nor would the members of parliament in Stockholm trust him to run it. In its reply, the court also made clear that Duhre received a certain amount of protection from criticism by being subjected to the estates. Having been granted the lease of Ultuna from parliament, Duhre was seen as a spokesperson for the *publicum* and as having submitted to the ultimate political authority. To argue, like Hanssen did, that Duhre had attained his position before others more qualified, was to criticise the judgement of the *publicum* and – as a result – to introduce disorder into the realm. The performances of Duhre and his *Laboratorium* were thus also part of a greater performance of the Swedish state.

Although Duhre won his case against Hanssen and the crofters, the controversies created by the court case undermined his position. The mere fact that he considered it necessary to present his side of this court case to the parliament of 1731 shows the lasting effects of the controversy. The court case also made him a less credible collaborator. He described:

> how these scattered evil rumours also have deterred several learned men from the candour and diligence, by which they [...] gladly had wished to serve the school at the Ultuna *Laboratorium* as its permanent members, and how I, because of this, have lost other indispensible people, such as craftsmen and other workers.[99]

The scholars' change of heart was not only the result of a general fear of moral contamination. The repeated irregularities undermined Duhre's ability to show that the *Laboratorium* would be permanent. He described how workers declined positions on his estate "by no other reason than that they have let themselves believe that the royal estate of Ultuna will be let out again, at the next parliamentary session, in the same way as before".[100] Because of these cases, in which Duhre's ability to perform as a master of his estate was put on trial, his audiences were starting to question whether the

---

förmån framför andra *meriterade*"; "Skrifwit håniskt och försmädeligt om de *œconomi*ske försök som H:r Duhre til Lands *œconomiens* underhielpande wälmenat *projecterat*"; ibid., 260.

[99] "huru dessa omkring spridda widriga ryckten, iemwäl hafwa afskräckt åtskilliga af lärda karlar ifrån den frimodighet och flit, hwar med de [...] gierna welat wara, til det erforderliga *Informations* wärckets tienst wid *Ultuna Laboratorium*, såsom des beständiga ledamöter och huru iag igenom detta kommit at missgå annat för mig omisteligt folk, af handtwärckare, och andra arbetare"; Duhre: "Memorial, 1731", 60–1.

[100] "af ingen annan orsak än at de låtit sig inbilla, at Ultuna Kungsladugård wid nestkommande Ricksdag kommer at bårtarenderas på samma sätt, som tillförende"; ibid., 62.

Laboratorium was permanent enough to be trusted. Therefore, it was essential for Duhre to perform successfully as a master of Ultuna in order to be recognised as a virtuous and credible partner among scholars, craftsmen and farmhands alike.

Cows and order in the King's Meadow

In the case against the crofters, Duhre had been accused of letting his own, and some of his friends', farm animals into the crofters' pastures, in a way that damaged the pastures in the long run.[101] Duhre was involved in yet another court case concerning the same issue, in which his position was undermined by the very farm economy that was supposed to support his *Laboratorium*. As mentioned earlier, after receiving the lease of Ultuna on 28 May 1724, Duhre applied for a lease of *Kungsängen*. He was not the only one interested in this pasture; the royal horseman [*hingstridaren*], Lars Kihlgren, had also shown an interest in the lease. In his application for the lease of the King's Meadow, Duhre argued that he would be a better tenant of the pasture. As part of his argument, he proposed new and harsher grazing rules to be imposed on him. He wished "that someone, through your lordship's authorisation, would have the power to [...] impose severe fines on him who causes more farm animals to be led into the pasture".[102]

At first glance, this proposition might seem strange. Why did Duhre want to impose harsh rules and heavy fines on himself? In the light of the responsibilities of a master of a royal estate, discussed earlier, it is easier to understand Duhre's actions. To submit to strict rules for grazing his farm animals was a way to be a virtuous and unselfish subject of the state. The orderly grazing of Duhre's cows projected Ultuna as a cameral model society, and its owner as a selfless mechanist–cameralist. Duhre's virtuous performance made sense because of the organic economy of grazing, in which meadows were considered a resource that needed to be carefully conserved, and grazing was an activity that needed to be balanced against the constraints of the land. To lead more farm animals than allowed into a pasture was a way for a tenant to earn extra money from a lease, but such additional profit deteriorated the grazing grounds in the long run. As the King's Meadow belonged to the crown, this was in effect a way of stealing from the state, or at least to put your personal gain ahead of the *publicum*. Consequently, if Duhre were perceived to be leading more animals onto the King's Meadow than the land could sustain, this would break his selfless performance. Such actions would present him as a man who had made

---

[101] Ibid., 238–9.
[102] "at någon af Eders Hög Grefl:e *Excell:ce* befulmäktigat måtte hafwa makt at [...] belägga den med ansseenlige böter, som är orsaken at flere Cratur blifwit insläpta uti ängen"; ibid., 274.

grand œconomical and mathematical promises to parliament, only to enrich himself using royal property.

Limiting the way he could use the grazing grounds was thus a way, as it were, to set the scene for future virtuous balancing acts in the King's Meadow. Duhre was successful in his proposal. On 30 June 1724, the Councillor of the Realm, Nicodemus Tessin, gave Duhre the lease of the King's Meadow and also accepted the new harsher rules.[103] However, some years later, Duhre's actions would drag him into yet another court case in which his *Laboratorium* and his character were at stake. On 18 June 1728, the royal horseman Kihlgren complained to the county governor's office that Duhre had led far more animals into the pasture than his contract allowed.[104] Duhre considered this case to be connected to the crofters' case and dismissed Kihlgren's accusations as "gross and intertwined falsehoods". Furthermore, he argued "that in time it will be possible to show that Kilgren [sic] never would have dared to present them, unless he, like my [...] crofters, had been enticed by a rumour spread by my opponents".[105]

Despite his strong denial of Kihlgren's claims, Duhre admitted that there had been more farm animals in the pasture than the contract allowed. There was a reason for this, however. He reflected on how "I saw myself placed between two difficulties". On the one hand, "in Uppsala I was described as one who grudges the inhabitants of the town of Uppsala the freedom to lead their farm animals into the King's Meadow of Uppsala, in exchange for the grazing fee". From the burghers' perspective, Duhre had admitted *too few* animals into the meadow. On the other hand, Kihlgren argued that Duhre had let *too many* animals onto the grazing grounds.[106] From Duhre's perspective, no matter how he acted, he could not please everyone. He considered himself caught in the middle of two conflicting interests, of which he could not satisfy both. Pleasing the locals in Uppsala required that he let them use the King's Meadow as they had done previously. But such a course of action would break the contract he had signed with the government in Stockholm. Duhre and his *Laboratorium* were stuck in a balancing act between the national and regional levels of eighteenth-century Sweden.

On 28 October 1730, the county sheriff Bohman carried out an inspection of the pasture, initiated by the county governor of Uppsala together with a group of local peasants. They found that the grazing cows had left the pasture in total chaos. Their report pointed out that "the grass roots are

---

[103] Ibid., 275–8.

[104] Ibid., 282.

[105] "så grofwa och sammanwridna osanningar"; "med tiden lärer kunna förmärkas, det Kilgren sig aldrig underståt med dem framkomma, så framt han äfwen som mina [...] torpare sig ey låtit förledas af et igienom owännen utsprit ryckte"; ibid., 291.

[106] "iag såg mig wara stad emellan twenne swårigheter"; "at iag i Upsala af owännerna uttyddes för den som miesunnar Upsala stads innewånare friheten, at i Upsala Kongs eng inläya sina creatur emot de wanl:e betespeng:ar"; ibid., 292–3.

partly pulled up by the farm animals and lay on the ground everywhere". The sheriff and the peasants also described how "the black soil is visible at some places and the Royal Stable's [*Kungliga Hovstallet*] pasture near the town, where the farm animals enter, is at places badly trampled and in bad shape".[107] The inspection did not look good for Duhre. He had, however, anticipated this outcome and initiated a parallel assessment of the meadow by the government in Stockholm. The county sheriff Petter Hartman, who had presented Duhre's fornication charges 2 years earlier, carried out this second inspection. The two inspections by Bohman and Hartman resulted in two completely different conclusions. In his report, Hartman argued: "that no parts of the King's Meadow of Uppsala, neither at the Royal Stable's nor at Ultuna's parts, has suffered any damage whatsoever".[108] Whereas Bohman and his group interpreted the roots lying on the ground as a sign of chaos, Hartman's group saw the orderly state of a field prone to flooding in autumn. Bohman argued that the grass roots lying on the ground were signs of destruction. Hartman, on the other hand, argued that the state was normal for pastures at this time of year. Whether or not the meadow had been preserved was thus not a simple matter of fact. The condition of the field was a matter of interpretation, and the resolution of whether the field was damaged or not was ultimately a matter of who carried the authority to define the situation. In the case of the King's Meadow, Duhre's inspection, ordered by the *publicum*, trumped the investigation initiated by the county governor. The King's Meadow was thus decided to be an orderly place.[109]

In many ways the differences between the two inspections corresponded to the evaluations of Duhre's *Laboratorium*. The performances of Duhre and of his cows were both subject to audiences' interpretations. The facts on the ground, the dirt and roots left where the cows had grazed, could be interpreted as signs of either chaos or order. Likewise, there did not exist a neutral matter of fact, a common ground or consensus, which made it evident whether or not Duhre had performed what he had promised the parliament of 1727. The interpretations of the trampled ground were thus one means for audiences in Uppsala and Stockholm, living in an agricultural society, to evaluate whether Duhre was a trustworthy and selfless mathematical–œconomical reformer of agricultural practice. The two inspections show that there was a rising ambivalence as to what was the actual intent of his project: public service or personal gain. In the case of Duhre's *Laboratorium*, the final judgement was made by the estates of the parliamentary session of 1731.

---

[107] "gresrötterne äro öfweralt til en dehl af creaturen upryckte ok på iorden liggandes"; "at swarta iuorden på några ställen synes och at Kongl. stal statens eng wid staden der creaturen hafwa sin ingång, är til en dehl illa trampat och medfaren."; ibid., 301.
[108] "at Upsala Kongeng å intet ställe, hwarcken å Kongl. Hof Stalletz eller å *Ultuna* dehlar har tagit någon den ringaste skada"; ibid., 310.
[109] Ibid., 309.

A morally ambiguous mechanicus in the parliament of 1731

During the parliament of 1731, the future of Duhre's *Laboratorium* was undecided. In preparation for the parliamentary session, Duhre had collected copies of documents concerning the controversies after 1727. Interestingly, he included documents concerning his cows and servants, but he did not include his fornication case in the proposal. Either he considered this case irrelevant to his work at Ultuna, or he chose to exclude it, perhaps because he was found guilty. Still, by binding his copies together into a handwritten folio and submitting it to the estates, Duhre himself judged the cases concerning his servants and his cows, at least, as relevant to the performance of his *Laboratorium*.

On 5 June 1731, the nobility discussed a report from the deputation now named the "Deputation of Cameral Matters, Œconomy and Commerce" [*Cammar- œconomie och commercie- Deputationen*] concerning the Ultuna estate. The report was not in favour of Duhre. After having heard the academic senate of Uppsala University, the deputation considered Duhre's mathematical lectures superfluous. Between 1727 and 1731, the university had appointed Duhre's former students Celsius and Klingenstierna professors – Klingenstierna of geometry (1728) and Celsius of astronomy (1730) – and they were now in charge of the mathematical education there.[110] Furthermore, according to the report, Duhre had failed to establish his *Laboratorium* on the estate; the deputation even considered this vision impossible to realise, given the resources at his disposal. Therefore, they wished to release him from his tenancy. Nevertheless, they believed that he had acted with good intent: it was out of "sincere love and out of the duty of a subject for the public good and enrichment", that he "had taken on more than has been in his power to bring about". Therefore, they suggested, Duhre should be considered free from all obligations to the public. The report also discussed the uncertain condition of the estate, noting that it had been argued against Duhre that he had "devastated the Ultuna forests", that he had "carried off hay", that he had "used the crofters [of Ultuna] on his own properties", and that he had "trampled the meadows". These aggravating circumstances, it concluded, should be evaluated in an inspection.[111]

The nobility reacted with anger to the deputation's consideration. Johan Fredrik Didron argued that "Duhre should be held responsible for having

---

[110] H. J. Heyman: "Anders Celsius", *SBL* 8 (Stockholm, 1929), 266; Sten Lindroth: "Samuel Klingenstierna", *SBL* 21 (1975–7), 319.

[111] "af uprichtig kiärlek och undersåtelig plicht för det allmennas upkomst och förkofring mera sig åtagit, än i dess krafter varit at efterkomma"; Sten Landahl: *Sveriges ridderskaps och adels riksdagsprotokoll från och med år 1719* vol. 6 [1731] (Stockholm, 1887), 423. "Ultunaskogens utödande"; "höets afförande"; "torparnes nyttiande på des egne lägenheter"; ibid., 422. "engarnes förtrampande"; ibid., 422–3.

deceived the members of parliament to give him the estate and [then] not to keep his promises." Similarly, Carl Henrik Wrangel was enraged that "Duhre has not kept what he has promised." Brauner – the new governor in Uppsala, who himself had handed in a counter-proposal to the deputation, was unsurprisingly the harshest critic. He described Duhre's shortcomings and argued that he instead should be the tenant of Ultuna. Duhre would "ruin the estate", were he to be left as its tenant. Also, Brauner insinuated economic wrongdoings: "I do not know if its commonly known that he has married and has given her 2,000 copper dalers, although he owns nothing and has donated everything he owns to the society."[112] Charles Emil Lewenhaupt calmed the members by pointing out that it would be unwise for the noble estate to "admit that they had let themselves be deceived", and that it would be best to let the king decide on the conditions by which Duhre would leave Ultuna. The rest of the nobility agreed with Lewenhaupt, and also decided that Brauner should be the new tenant of Ultuna.[113]

On 11 June, the same report was read to the peasants, the clergy and the burghers. All estates agreed with the nobility; however, the peasants stressed that Duhre should be responsible for the damage he had done to the fields and meadows.[114] In the parliament of 1731, the positive view of Duhre as a virtuous and useful man had thus all but disappeared. While the deputation did not consider him to have deceived the *publicum* intentionally, he was no longer deemed a competent *mathematicus* who could raise the realm from poverty singlehandedly. Instead, they viewed him as a man who did not know his own limitations. The members of the noble estate were even less favourable of Duhre's person. For them, he was an imposter and a con man. The only reason for not scolding Duhre publicly was the reluctance of the estate to admit to having been deceived in the first place.

Interestingly, many of the local controversies that unfolded between 1727 and 1731 are mentioned in the protocols of the estates: Duhre's relationship with his crofters, his handling of the fields and his personal relationships. Whereas his successful performance as a good *mechanicus* in 1723 and 1727 was interwoven with the presentation of a moral man, his loss of Ultuna in 1731 was likewise linked to his character. The former good

---

[112] "*Duhren* borde vara ansvarig för det han bedragit Riksens Ständer till at lemna honom kongsgården och icke hållit det han lofvat."; Landahl: *Sveriges ridderskaps och adels riksdagsprotokoll från och med år 1719* vol. 6 [1731], 423. "at *Duhren* ey hållit hvad han lofvat."; ibid., 424. "så ruinerar han gården"; "Jag vet intet, om det är bekant, at han gift sig och har gifvit henne tvåtusende Dahl. kmt, them han likväl intet äger, uthan förskrifvit alt thet han äger till Societeten."; ibid., 423. On 4 March 1731, Duhre had married the widow Ingeborg Stefansdotter, previously married to the peasant Erik Pedersson. Bengt Hildebrand: "Anders Gabriel Duhre", 506. This timely marriage could of course be related to lacking third power relation, between a husband and a wife, at Duhre's estate.
[113] "at säija det de hafva låtit bedraga sig"; ibid., 424.
[114] Sten Landahl: *Bondeståndets riksdagsprotokoll* vol. 2 [1731–4] (Stockholm, 1945), 128; Norberg: *Prästeståndets riksdagsprotokoll* vol. 7 [1726–31], 664–7; Nils Staf: *Borgarståndets riksdagsprotokoll från frihetstidens början* vol. 4 [1731] (Stockholm, 1958), 187.

mechanicus was now received as an immoral man who deceived the public for personal gain, or at least as an overly optimistic man who did not know the limits of his powers. As a junior civil servant in Stockholm, Duhre had been a pious part of the hierarchical bureaucracy. There, it had been clear who was his superior, and, consequently, what expectations his audience had of his performance. In the parliament of 1722, he had been required to appease a more heterogeneous audience, but his performance was still a straightforward affair of presenting a vision of the future in printed form. At Ultuna, conditions had been different. As I have shown, in order to understand his performance, one has to see Duhre and the Ultuna estate as a team performance. Success or failure depended on how audiences in Stockholm and Uppsala perceived the master–servant relationship at Ultuna and how it related to other local groups. Whether or not Duhre was convicted in the court cases concerning Ultuna, they conjured forth an alternative image of the estate. Instead of being a vision of a good cameral order, founded on a mathematical–œconomical foundation, it had become the mirror image of a disorderly state. According to this image, Duhre failed as a master because of his inability to control his violent and selfish impulses. Was Duhre selfless and virtuous, or moody and violent?

Epilogue. Œconomical dreams of a bad mechanicus

By the late 1730s, Anders Gabriel Duhre's reputation had hit rock bottom. In the parliamentary debates, his project was generally considered a failure. And worse: Duhre was perceived as a man who had promised the world but had failed to deliver. He lost the estate at Ultuna, and was publicly denounced as an imposter and a con man. Destitute and in debt, he moved back to his family estate at Gnista, inherited by his older brother Carl after the death of their father in 1726. But Duhre did not accept his defeat. He submitted to the parliament of 1738–9 an extensive handwritten proposal similar to the one he had compiled in 1731.[115] Also, he started writing what would become his last work: *The highest wealth of Sweden, built on an œconomical foundation* (1738). This work was Duhre's last attempt to convince his contemporaries of the merit of his mathematical and œconomical dreams. It was also where he made his boldest and most comprehensive claims. There, he presented a whole new ideal state, organised after *œconomical* and mathematical principles.[116]

---

[115] Anders Gabriel Duhre: "Handlingar ang. Ultuna. Ingivna av Anders Gabriel Duhre", 1738–9, Frihetstidens utskottshandlingar, R 2702, RA.
[116] Anders Gabriel Duhre: *Sweriges högsta wälstånd, bygdt uppå en oeconomisk grundwal* (Stockholm, 1738). The same year, Duhre also published an excerpt of the same work; Anders Gabriel Duhre: *Utdrag af Anders Gabriel Duhres oeconomiske tractat, kallad Sweriges högsta wälstånd, hwarutinnan*

His argument was structured in the form of a geometric proof, starting from three principles. First: that virtue was the same as the contribution to "the edification and multiplication of human coexistence". Second: that the habits people acquire when young set them on a way of life, which they only with difficulty can turn away from. Third: public usefulness should be considered the principal law of the realm. Following this law, everyone should only conduct work that they are capable of, and anyone who does not contribute to the common good should be considered "pernicious vermin".[117] Therefore, in Duhre's ideal realm, the number of scholars should be limited. No more people should lead this life than the number that "is proven to be indispensible for the filling of the required offices and positions in the clerical and the worldly estate."[118]

Instead, his ideal realm was to be filled by men "who care for necessary knowledge in *Mathematics* and [...] *physics* gained through experiments". These men should first have "acquired a mature skill and experience in techniques, crafts and all other practice". Then they would "apply these principle sciences [i.e., mathematics and physics] to the affairs of the realm."[119] Hence, Duhre had sharpened, rather than changed, his rhetoric as a consequence of his own radically altered conditions. He still based his argument on the promised fruits of virtuous men raised through mathematical and artisanal exercises. The œconomical principles in *The highest wealth of Sweden* are thus not very different from those found in Duhre's earlier works. Although he proposed changes at a whole new level in this work, changes that would fundamentally change the social order, these reforms were motivated using the same arguments found in his works from the 1720s.

The reception of this work, however, was the complete opposite of that of his publications of 1722. The bad response can hardly be attributed to a diminishing interest in œconomical works in the 1730s and 1740s. These decades saw a number of prominent works that were similar to Duhre's.[120]

---

sammanhanget så wäl som sakens höga wigt och angelägenhet, til alla wälsinte patrioters widare ompröfwande, warder kort och tydeligen å:daga lagd (Stockholm, 1738).

[117] "mennskliga sammanlefnadens uppbyggelse och förmering"; Duhre: *Sweriges högsta wälstånd*, 6. "fördärfwelig ohyra"; ibid., 7–8.

[118] "pröfwes oumgängelige til nödig tiensters och bestälningars beklädande uti det Andeliga och Werldsliga Ståndet."; Duhre: *Sweriges högsta wälstånd*, 9.

[119] "som winlägga sig om en erforderlig kundskap uti *Mathematique* och den der uppå grundade Natur kunnigheten, förwärfwad igenom försök"; "förwärfwat sig en mogen snäll- och erfarenhet uti handalag, handwerken och al annan *practique*"; "lämpa dessa Hufwud-Wetenskaper til Landets angelägenheter."; ibid.

[120] Duhre's work can be compared to the contemporary œconomical works by Anders Nordencrantz: *Arcana oeconomiæ et commercii, eller Handelens och hushåldnings-wärkets hemligheter* (Stockholm, 1730); Olof Hamrén: *Manufacturs-spegel, som genom tilförlåtelige uträkningar och uprichtiga grundsatzer gifwer et fulkommeligit lius om manufacturers och fabriquers rätta art och beskaffenhet, samt oskattbara nytta genom försiktig anlägning* (Stockholm, 1738); Lars Salvius: *Tanckar öfwer den swenska oeconomien igenom samtal yttrade [...] emellan fru Swea, fru Oeconomia, Herr Mentor, Herr Flit, Herr Sparsam, och Frans.* (Stockholm, 1738); Carl Linnaeus: "Tankar om grunden til œconomien genom naturkunnigheten

What had changed though, was Duhre's reputation. Whereas the members of the parliament of 1723 viewed him as the embodiment of a mechanicus, 15 years later his character did not carry the same sway. His performance in Ultuna had compromised his trustworthiness in the eyes of his contemporaries. Thus, when he revealed that his work at Ultuna had only been a model that showed how society as a whole should be organised, many were sceptical of his promises. After the failure of his *Laboratorium*, they required more from him than the grand promises of a publication on œconomy.

One reaction to Duhre's last work was a satirical poem in Latin. It presented the "famous" Duhre and his work:

A name since long remembered in the hyperborean realms,
Duhre, also in our patria a famous old man
Hitherto secret, a long time in the making
Lo! To your eyes, patria, he presents a work
No other parent, has been able to conceive as worthy an
offspring: every page speaks of its master.[121]

Although the anonymous author was never explicitly critical of Duhre, the poem magnified his promises to the absurd. The poem described how Duhre's work would bring amounts of gold and riches to the realm, larger than those that had been brought back from the New World; how his work would let the fields and craftsmen of Sweden feed millions; and how "the hyperborean lands" would come to attract brilliant minds from abroad. Duhre's work was "the stone, for a long time sought for in vain/which truly will give us a golden age".[122] *The highest wealth of Sweden* magnified Duhre's own claims from the 1720s to a degree that might have been considered absurd to his contemporaries, given the failure of his *Laboratorium*. This poem suitably magnified the praise and promises of his first proposals, so as to make his new claims seem even more bizarre.

---

och physiquen", *Kungliga vetenskapsakademins handlingar* (1739–40), 411–29; Jonas Alströmer: *Sveriges wälstånd om det will* (Stockholm, 1745).
[121] "Nomen hyperboreis dudum memorabile terris, | DUHRIUS, & patrio notus in orbe senex, | Hactenus occultum, meditatum tempore longo, [En! tibi conspicuum, patria, sistit opus. | Haud alio nasci potuit tam digna parente | Progenies: dominum pagina quæque refert."; Anonymous: *In opus Andreæ Gabrielis Duhre, summa felicitas Sveciæ, dictum, Svetice, Sueriges högsta wälstånd.* (No place, undated); See also Grape: *Något om Anders Gabriel Duhre och en honom ägnad latinsk dikt.* There also exists a shorter version of the poem, printed 1738 in Stockholm, titled *In Andreæ Gabrielis Duhre opusculum svethicum, dictum Summa Sveciæ felicitas, Svethicè, Sweriges högsta wälstånd* (Stockholm, 1738).
[122] "Hic lapis est, longo frustra quæsitus ab ævo, / Qui vere nobis aurea secla dabit."; Anonymous: *In opus Andreæ Gabrielis Duhre, summa felicitas Sveciæ, dictum, Svetice, Sueriges högsta wälstånd.*

Conclusions. Integrating the good and the bad mechanicus

The history presented in this chapter, following Duhre from his parents' estate at Gnista and back again, is hard to reconcile with a heroic narrative. While he presented dreams of prosperity and order, his history is one of making and unmaking. Therefore, it is not possible to characterise him as a certain kind of man – good or bad, competent or useless – using the available historical sources. Even in the decades following his death, Duhre was discussed in numerous and diverse ways. In a speech at the Royal Swedish Academy of Sciences, commemorating Ekström, Wargentin called Duhre "laudable" and his "*Laboratorium Mechanicum* [sic]" a school "established with very good intentions". At this *Laboratorium*, Duhre had hired "a German born erudite *Mechanicus*, named Fuchs" but the academy came to a halt, because of "jealousy and persecution".[123]

In a similar speech about Celsius, von Höpken presented Duhre's school as a shelter where Celsius "could protect himself under the permit, that the late Gabriel Duhre [sic] had received from parliament, to teach the youths at the university."[124] Similarly in 1768, commemorating Klingenstierna, Strömer told of how Duhre "was a famous mathematicus at that time [in 1718], and an ardent promoter of such studies". But, according to Strömer, Duhre "was not much of a speaker, expressed what he thought unclearly, and was usually so confused that he could not see what happened in the company of others."[125] Similarly, in Torbern Bergman's speech about Brandt, Duhre was described as someone who was very competent in mathematics (even "more than at that time was considered useful or advisable"), but also "somewhat strange in his social intercourse, because of a lack of good upbringing".[126] Most early twentieth-century historians of Duhre picked up on Bergman's and Strömer's negative descriptions. In a biographic article on Duhre, Hildebrand managed to summarise both the positive and negative portrayals of him in a single sentence: "despite all his noticeable flaws, the enthusiastic Duhre was a pioneer for many modern ideas, especially as regards the technical vocational education and the rationalisa-

---

[123] "berömvärde"; "i mycken välmening upprättad"; "en til börden Tysk vitter *Mechanicus*, vid namn FUCHS"; "genom afund och förföljelser"; Wargentin: *Åminnelse-tal öfver Daniel Ekström*, 11.
[124] "måste han skydda sig under det tilstånd, framledne GABRIEL DUHRE af Riksens Högloflige Ständer erhållit, at vid Högskolen få undervisa ungdomen."; Höpken: *Åminnelse-tal öfver [...] Anders Celsius*, 11.
[125] "en berömd Mathematicus, på den tiden, och nitisk upmuntrare til det studium"; römer: *Åminnelse-tal öfver Samuel Klingenstjerna*, 12. "hade inga talegåfvor, kom oredigt fram med det han tänkte, och var oftast så tankespridd, at han icke såg hvad som hades för händer i sällskaper."; ibid., 12–13.
[126] "längre, än i allmänhet den tiden räknades både nyttigt och rådligt"; "något sällsam i omgänge, af brist på tjänlig upfostran"; Torbern Bergman: *Åminnelse-tal öfver framledne bergs-rådet och medicinæ doctoren ... Georg Brandt, hållet för kongl. vetenskaps academien, uti stora riddarehus-salen, den 9 aug. 1769* (Stockholm, 1769), 7.

tion of farming."[127] Many twentieth-century histories of Duhre's *Laboratorium* describe these events as a personal failure that can be explained by Duhre's strange character and lack of social skills, a character they infer from the short comments in Strömer's speech. That is, they see the events as an effect of tragic inherent traits in Duhre himself.[128]

Such a psychological explanation is hard to reconcile with Duhre's life up until 1727 (i.e., his first 46 years). As seen in this chapter, during those years Duhre made a career in the state bureaucracy in Stockholm, published important mathematical textbooks and, ultimately, charmed the Swedish parliament into giving him a royal estate. Up until this point, few sources suggest that he had a hard time making his thoughts clear or that he could not handle himself socially. In sources later than 1727 though, he comes across as a less successful and selfless man. Rather than to suggest that he had a sudden change of personality during the autumn of 1727, his shifting character provides an opportunity to contemplate how historical actors present themselves to contemporary audiences and posterity.

Like Polhammar/Polhem in the previous chapter, Duhre was hardly a static person. Instead, he was constantly occupied with making and remaking himself through mathematical and mechanical techniques. Also, as in the case of Polhem, Duhre ended up in a relationship with the *publicum* that was not positive for his career or his reputation. While Polhammar nourished a close relationship with Karl XII and the absolutist regime, which was not helpful after the death of Karl XII, Duhre projected an image of unconditional love for the *publicum*, which involved promising results that were hard or impossible to deliver. Whereas Polhammar's relationship with the absolutist king became a problem for him during the early constitutional monarchy, Duhre's relationship with the estates became a problem in relation to the social framework of the local agricultural community around Uppsala. The complexities of performing as a master of an estate reflected badly on Duhre and his mathematical–œconomical project, and this failure contributed the low support for his project in the parliament of 1731. At the heart of Duhre's *Laboratorium* lay a vision of an easy fix for the Swedish state's serious economic problems. Therefore, Duhre needed to promise

---

[127] "Med alla sina påtagliga brister var den entusiastiske D. en föregångsman för många moderna idéer, särskilt rörande den tekniska yrkesutbildningen och lantbrukets rationalisering."; Hildebrand: "A G Duhre".

[128] These characterisations are prevalent through much of the studies on Duhre from the first half of the twentieth century. Duhre is described as: "a rather strange person" (in original: "en ganska egendomlig personlighet"); Henrik Schück: *Ansatser till ett universitet i Stockholm före 1800-talet* (Stockholm, 1942), 39. Furthermore, he has been described as "a great sanguinian, an enthusiast who did not take the realities of life into account" (in original: "en stor sangviniker, en entusiast som föga räknade med livets realiteter"); Hebbe: "Anders Gabriel Duhres 'Laboratorium mathematico-oeconomicum'. Ett bidrag till Ultunas äldre historia", 590. Likewise, Duhre's ideas have been described as "immensely naïve" (in original: "gränslöst naiva") as well as "original – and yet uninteresting" (in original: "originella – och likväl ointressanta"); Schauman: *Studier i frihetstidens nationalekonomiska litteratur*, 115–16.

much, and constantly raise the stakes, in order to appease his audiences. These raised stakes meant that the complexity of Duhre's performances increased over time: he came to rely on an increasingly heterogeneous set of supporting actors, in the community surrounding his estate, and, likewise, the audiences of his performances multiplied.

Duhre's performances of the good mechanicus carried the seed of his future failure. In order to succeed in his performance of the useful mathematical–œconomical subject in 1723, he needed to arrange the scene and raise expectations to the extent that he could hardly avoid performing as a bad mechanicus some years later. Some contemporary audiences, such as the Deputation of Cameral Matters, Œconomy and Commerce of 1731, were more aware of this unstable dynamic than most twentieth-century historians. They saw that Duhre had promised too much in his earlier presentations of his *Laboratorium*, but that he had done this out of love for the *publicum*. In their eyes, this freed him of guilt. He would never have been able to deliver what he had promised the optimistic parliamentarians of the 1720s: singlehandedly to turn the Swedish state into a well-ordered society. For some, Duhre's actions was good enough, as long as it was perceived that he had held a virtuous intent when he had courted the *publicum* with his impossible mathematical and œconomical dreams.

6. Conclusion. Performing the mechanicus

This book is about mathematics and mechanics in the early modern Swedish state. However, unlike most previous work on the subject, my focus has not been the history of ideas, or the practices of machine building and knowledge making. Instead, I have studied mechanics as an exercise of virtue, which nurtured a certain kind of man. This man was denoted by the Latin word "mechanicus", roughly translatable as "a man of mechanics". The mechanicus was a persona, existing in between the individual and the collective. This persona consisted of a number of expectations as to who a mechanical practitioner should be. There was an expected life narrative, stipulating how a mechanically apt boy should be raised into a man. This man, in turn, was expected to possess certain skills, and he was supposed to perform them in a specific way. Last, but not least, in his performance of mechanics, he was to relate to his superiors and subordinates in a way that identified him as a pious subject.

The mechanicus, the object of study of this thesis, was a performative actors' category. In other words, the expectations that constituted it existed in the encounters between mechanical practitioners and their audiences. Consequently, I have approached the mechanicus through the numerous relationships by which mechanical practitioners made themselves relevant to their time. Such relationships are found in the coming-of-age narratives of commemorative speeches and textbook prefaces, in the community of men working in the Bureau of Mines in Stockholm, in the correspondence between Polhammar and the absolute monarch Karl XII, and in Duhre's efforts to reform the Swedish constitutional monarchy of the 1720s by mathematical, œconomical and mechanical means. These relationships were conduits of intertwined dialectic processes, which not only moulded a man into the form that was expected of a mechanicus, but also changed actors' and audiences' expectations of how a mechanicus should act. In other words, mechanical practitioners were part of an early modern social epistemology, which was both flexible and dynamic.

Here I develop and draw together the main observations of this thesis. My conclusions are presented in three thematic sections, each discussing one important aspect of my study. First, I discuss the implications of seeing the mechanicus as a persona, formed in relation to the political order of the early modern Swedish state. Second, I consider how studying mechanics as relational performances facilitates new means of understanding the diverse and shifting meanings early modern actors ascribed to it. Finally, I analyse the changes in the role of mathematical and mechanical performances between 1700 and 1750, and suggest that we should approach these changes by focusing on historical actors' own understanding of them.

The mechanicus of the Swedish state

The mechanicus of the early modern Swedish state lay claim to a socio-epistemic space, defined in relation to a vast number of early modern groups and actors. This heterogeneous group of relational others consisted of, for example, academic scholars, craftsmen, pure mathematicians, common miners, monarchs, peasants, the sovereign and even God. The expectations that constituted the mechanicus can be seen as a dialectic of difference and similarity: he was defined by relationships of superiority and subordination to all of these other actors, as well as by distinctions of gender, age and social status.

But then who was the mechanicus? First, he was a man of mature age. To identify the mechanicus as a man is not merely to recognise a simple matter of fact, but also to suggest that the practices, social structures and imaginations of mechanics carried certain expectations on how to perform gender. The mechanicus was imagined as having once been a boy, who had entered useful manhood through the exercise of mathematics and crafts. These exercises had made him embrace certain virtues, such as concord and diligence, as well as the *epistemic* virtues *theoria* and *praxis* (i.e., codes of performing both as a pious subject and a trusted knowledge maker).

Whereas pure mathematics, and especially geometry, was recognised as an exercise of discernment and concord, the exercise of crafts nurtured a diligent and hardworking manly subject. When integrated, the exercise of mathematics and crafts were seen as nurturing the epistemic virtues of *theoria* and *praxis* in a boy. Because the mechanicus was identified as possessing this broad range of manly virtues, he was also recognised as a useful man who could make important contributions to the body politic. This virtuous and useful man performed selflessly in relation to his contemporaries, and his actions were directed by the needs of the *publicum*. My observations thus reinforce Runefelt's argument that, in early modern Sweden, the concepts of virtue and usefulness were interchangeable. However, having studied virtue and usefulness as relational concepts enacted in performances, I have shown something more than how these terms were related in early modern publications. Early modern men also performed virtue and usefulness in relation to their contemporaries, and a man was required to display proper submission and superiority, in a number of relationships, in order to be virtuous and useful. Usefulness and virtue were closely related because they were both means of performing as a good subject, aligned with the political order. Mechanics was useful because it integrated pure mathematics with crafts. By balancing the connotations of these two sets of techniques, and by presenting their combination as novel, mechanical practitioners presented themselves as men who were able to rearrange, and to perfect, the relationships of the early modern state according to the demands of the sovereign. Thus, the usefulness of mechanics cannot be separated from the way in which it shaped a virtuous man. This man, the mechanicus, could amend the imbalances of the early modern Swedish theocracy in order to preserve it, while his machines, consisting of correctly placed cogwheels, corresponded to a harmonious state and an orderly nature.

When a mechanical practitioner succeeded in performing as a mechanicus, he was recognised as a useful and virtuous man. The virtues of mathematics and crafts structured the intergenerational relationship imagined in textbooks, and they formed the basis for an emerging symbolic role of geometry and mechanics in the Bureau of Mines of the first half of the eighteenth century. They were even constitutive of men, as in the case of Polhammar, deemed worthy as clients of absolute monarchs, who in turn shaped themselves into enlightened sovereigns through patronage of mechanical work. In the constitutional monarchy, as the case of Duhre shows, such a man became a symbol, uniting parliament around a vision of an orderly future. However, while eighteenth-century audiences associated certain characteristics with a good mechanicus – virtuous conduct, disinterestedness and submission to the ultimate power – they also knew how to identify a bad mechanicus. Failure to meet the expectations of how to perform as a mechanicus brought disgrace not only to the practitioner himself, but also to the very techniques that he performed. This antithesis of the good mechanicus was identified by his lack of self-knowledge, his false promises, his appetite for private gain and his useless inventions. A bad mechanicus was either an accomplished but selfish man, or someone who claimed to possess skills that he did not have. In either case, such a man used the positive connotations of mechanics for his own personal gain to the detriment of the common good. The grand promises of mechanical practitioners always risked being interpreted as the claims of imposters who promised too much, or who sought to benefit themselves. Thus, the performances of the virtuous mechanicus also carried the seed of the bad mechanicus.

Early modern audiences had seemingly contradictory expectations of a mechanicus. He was a free actor, who could be compared to the sovereign, or even God, while at the same time he was recognised as a lowly civil servant who could be trusted because he was integrated into the strict hierarchies of the early modern state. In the patriarchal Lutheran state of Sweden, where relationships of superiority and subordination were never a zero-sum game, acts of submission could be enabling. In this hierarchical state, where private economic resources were scarce, the act of submitting – either to one's elders, to the state apparatus, directly to the sovereign, or to the nebulous *publicum* of the constitutional monarchy – was a way to become trusted to act. Because submission was a prerequisite for action, Swedish mechanical practitioners never possessed "free action" of the kind that Shapin describes as central to English seventeenth-century gentlemen. In the English "great civility", all "free action had to be freely disciplined".[1] In the Swedish state, however, men were part of a web of asymmetric relationships, ultimately leading up to the sovereign or the *publicum*. To be trusted was to assume an easily identifiable, and stable, position in this web, and to perform mechanics was one way to claim such a position. Also, performances of mechanics were interpreted differently depending on the position of the subject who carried them out. What this person had done previously – his social status and his character – mattered to how his mechanical performances were received by audiences in the Swedish state.

Paradoxically, mechanics was imagined both as an activity that transgressed socio-epistemic categories and as an exercise that fostered a submissive, and obedient, subject. For many mechanical practitioners, this paradoxical position was productive. On the one hand, these men highlighted the creative ways in which they transcended imagined socio-epistemic dichotomies. That is, they could present themselves as existing in between, and therefore outside, opposing concepts and groups such as those of *theoria* and *praxis*, or mathematicians and scholars. On the other hand, these socio-epistemic transgressions were not seen as threats, as long as the mechanicus was considered to act in the interest of the *publicum*. Because of this productive paradox, the creative transgressions of the mechanicus were feasible means of self-presentation. Because sovereigns, and other patrons, expected a mechanicus to transgress these boundaries, historical actors did not consider these transgressions as challenges to the political or religious order. Instead, they were *expected transgressions*, aligned to the theocratic order of the early modern state.

Because these transgressions were expected, it is important to be careful when interpreting autobiographical writings by early modern mechanical practitioners. Such sources are more than neutral self-descriptions: they were formulated in relation to certain audiences with specific expectations. In many cases, documents, which at first glance might appear to be written by historical actors existing in between contemporary cultures, were actually produced by actors who catered to the dreams of a balanced state of their audiences in government and in the cameral bureaus. Recently, historians of science and technology have begun to study the role of "go-betweens" and "hybrids" in knowledge making.[2] My cases underscore how one should be cautious when using such modern categories to describe historical phenomena. At least, it is important to specify exactly what we mean when discuss-

---

[1] Shapin: *A social history of truth*, 39.
[2] For examples of such studies, see page 20 note 23.

ing early modern engineers and mechanical practitioners in these terms. Can we really understand this category of men as truly transgressive if contemporary audiences repeatedly expected, or even demanded, that these transgressions be made? Do not such categorisations risk becoming a new form of celebratory history, which misses the actual roles that these men had in early modern cultures?

Relational performances

By studying the performances of mechanical practitioners, I set out to understand the mechanicus as more than an ethereal ideal. Performances of mechanics were made in various relationships of numerous types and between heterogeneous actors. However, these relationships shared some characteristics. First, they were *asymmetric*. In other words, they were not relationships of peers, but between actors defined by different social distinctions such as gender, social status and age. Second, they were *mutually shaping*. Not only did the relationships change those who entered into them, they also formed what was expected of a mechanicus. Third, the relational performances of mechanics were *balancing acts*. That is, they needed to be repeated continuously and their outcome was never certain. A man did not become a mechanicus once and for all: actors constantly needed to relate and adapt to the changing expectations of their audiences.

The asymmetry of early modern relationships has been pointed out by numerous studies of early modern patron–client relationships. These studies have generally focused on clients' tactics to attain authority and resources for their work, as well as on patronage as a means to gain symbolic capital.[3] Similarly, histories of science often portray historical actors as rational agents who used certain premeditated strategies in order to reach clearly defined goals: such as making knowledge, building instruments or asserting authority. I have wanted to nuance such accounts by showing how these asymmetric relationships profoundly shaped the goals and priorities of the men who were part of them. Ultimately, these relationships had a performative dynamics of their own, beyond the control of either of the involved parties. What started out as a mutually beneficial relationship could in time develop into an affiliation that proved problematic for both parties.

Such relationships were more than alliances of necessity. For example, while the initial relationship between Polhammar and Karl XII can certainly be described as a mutually beneficial one between a client and a patron, Polhammar's emotional connection to the monarch after 1718, which was certainly not beneficial either to him or the dead king, is more difficult to understand from such a perspective. Polhammar's loyalty to the late mon-

---

[3] For examples of such studies of patronage, see page 125-6 note 9.

arch was the result of a mutually shaping relationship based on sameness and difference. In such a relationship, the partners partly surrendered control of who they were. The newly ennobled Polhem of 1720 was thus shaped by, and entangled in, a relationship that was no longer favourable to him. Leaving the relationship, however, meant abandoning a part of whom he had become, and the only way to shed parts of oneself was to engage in new relational performances, made for other audiences. Thus, Polhem would not leave Karl XII fully behind until he found a new audience in the Royal Swedish Academy of Sciences of the mid-1700s.

The actors who entered into these mutually shaping relationships struggled with how to relate to one another without losing themselves. On the one hand, men who wanted to shape themselves in each other's image needed to become emotionally close. On the other hand, emotional connections always involved submission. No matter if you were the superior or subordinate party, to engage in an asymmetric relationship was both enabling and disabling: you always gave up some means of agency in order to gain others. For example, the imagined intergenerational relationships between mechanically apt boys and experienced mechanical practitioners involved close and loving emotions. These cordial emotions, however, needed to be held in check, so that the parties to the asymmetric relationship did not become peers. The uncertain emotions of these relationships made them unstable. They were structured around a love that rose from mutual recognition, but it came to a halt as soon as the involved parties became too similar. Because these relationships always required both intimacy and detachment, they were constant balancing acts.

Often, the relational performances of mechanics cannot simply be understood as a relationship between an actor and an audience. For example, the Bureau of Mines was a community of men consisting of a web of relational performances. Its officials simultaneously performed for each other, and perceived the performances of their colleagues. Also, many relational performances of mechanics were re-enacted by a larger circle of actors for wider audiences. The relationship of Karl XII and Polhammar included correspondence, consecutive letters that each moulded the two men in slightly new forms. Thus, like the Bureau of Mines, this chain of correspondence cannot simply be described as a series of self-presentations made by *one* actor in relation to *one* audience. In this correspondence, the actual individuals behind the names "Karl XII" and "Polhammar" were less important than the symbols that they constituted. For example, the king's own voice was seldom heard in these letters. Instead, his reactions, opinions and wishes were described, interpreted and forwarded by a team of men making themselves spokespersons for the monarch. In the end, it is not even certain that he was personally involved in his relationship with Polhammar.

Identificatory acts were not only made in relation to immediate performances, but also in relation to the perceived character of the actors who performed them. Whether or not contemporary audiences identified a per-formance of mechanics as successful depended partly on the relational his-tory between the mechanical practitioner and the given audience. On the one hand, the logic of these relationships was straightforward: a mechanical practitioner who had been deemed successful in previous projects was likely to be trusted again. On the other hand, success also generated negative feedback: previous success tended to raise audiences' expectations, and me-chanical practitioners thus constantly needed to raise the stakes of their performances. Such negative feedback can be seen in the rise and fall of Duhre and his *Laboratorium*. Duhre's initial success, in the state bureaucracy and as a mathematical author, reflected well on his character and made it possible for him to acquire the lease of Ultuna. At the same time, the very same success multiplied his audiences and raised their expectations of what a mechanicus, such as him, could accomplish. In the end, he was unable to meet the increasingly high demands of his diverse audiences and, conse-quently, he and his mathematical visions fell from grace. Similarly, failure to perform did not necessarily have immediate negative effects for an actor. Often audiences, invested in specific relational performances, had an incen-tive to cover up failed performances, at least to a degree. Admitting that an actor had failed in his performance could reflect badly on the audience, who had shaped itself in relation to a mechanical practitioner. Perhaps the clear-est example of such a cover-up is the discussion about Duhre's project in the noble estate of the parliament of 1731.

Early modern dreams of mechanical perfection seldom, if ever, came to fruition. The Swedish state, although imagined otherwise, was a disorderly and ever-changing structure. Likewise, the transgressions expected of a mechanicus were unstable as well. I have approached the line between the good and the bad mechanicus, or between success and failure, not as de-fined by essential qualities of men, but as drawn by actors' performances. In other words, audiences identified character traits in an actor, such as virtue and usefulness, according to how he performed; a man who might have performed virtuously at one time could soon lose this trait in the eyes of his audience. Therefore, if we want to fully understand the role of the mechani-cus of the early modern Swedish state, we have to study how mechanical techniques were performed in relationships between actors and audiences. By studying such relationships, I have highlighted the diverse processes by which a mechanicus was made and unmade through anxious mechanical performances.

Taking the "modern" out of the early modern

I have wanted to disentangle eighteenth-century mechanical practitioners from long narratives of the rise of industrialism and the making of modernity. The actors studied in this thesis differed from the entrepreneurial individuals identified by economic historians and historians of technology. By relating the performances of these actors to the expectations of their contemporaries, rather than to long-term effects, I have shown how the historical actors themselves perceived the merits and risks of being a mechanicus, and how they performed in relation to this persona.

Although this has been a synchronic study, not primarily interested in how the role of the mechanicus changed over the studied time period, it has nonetheless identified some interesting changes between 1700 and 1750. As visible in the applications to auscultate in the Bureau of Mines, as well as in mathematical publication by auscultators and in documents that they transcribed (all discussed in Chapter 3), the role of mathematics in the Swedish mining administration – and especially of geometry and mechanics – changed during the first half of the eighteenth century. In many ways, this development is the opposite of what would be expected from a rising technocratic management of a mechanised and industrialised mining production. While, in the early eighteenth century, mathematics (in the broad early modern sense) was almost exclusively linked to mining mechanics, by the 1720s, it had gained a new symbolic role of relevance to a wider group of mining officials. Whereas historians such as Mokyr and Jacob have argued that the application of mathematical prescriptive knowledge to the know-how of craftsmen was significant to the industrial revolution, it seems as if, in the Bureau of Mines, such "application" predated a more general use of mathematics. Similarly, while historians have argued that the mathematisation of crafts was part of a process of disciplining craftsmen, in the Bureau it seems as if mathematical approaches spread from mechanical practitioners, who exercised crafts, to a broader group of cameralists. If we were to use the Bureau as an example of broader developments, we would thus end up with a model of how early modern mathematics, which at the beginning of the 1700s was recognised as an applied form of knowledge used for building machines, by the mid-eighteenth century also gained a symbolical role in defining a distinct category of cameral men.

This development of a symbolical role of mathematics can also seen when comparing the careers of Duhre and Polhammar. Whereas Polhammar's career during the Swedish absolutism was primarily based on his being identified as a man who could save the realm through his machine building, Duhre was mainly seen as a similar saviour because of his ability to teach mathematics and to foster boys into virtuous and useful men. Although there was no radical break here – Polhammar had an interest in mathematical and mechanical education, and Duhre dabbled in machine building – their self-presentations still leaned in different directions. This difference is further accentuated if we see both these men as products of the community of the Bureau of Mines. It is thus possible to see them as embodying the different roles of mathematics in the Bureau of the early and mid-eighteenth century. Along with the changed roles of mathematics in the early modern Swedish state, the audiences for mathematical and mechanical performances also changed. When the government was radically transformed from an absolute monarchy to a constitutional one, the personal relational performances between the monarch and mechanical practitioners were replaced by performances for a much broader audience. This change in audiences went hand in hand with the new role of mathematics at this time. For Duhre, to link mathematics to a much broader set of activities, and not only to mining mechanics, was a way to show his relevance to his heterogeneous audience.

How are we to understand the changing role of mathematics and the emergence of broader audiences during the first half of the eighteenth century? A key argument of this thesis has been that early modern actors and audiences considered mechanics meaningful *because* they shared anticipations of how a mechanical practitioner should act. This argument might appear self-evident. However, as discussed in Chapter 1, a number of historians, following a liberal narrative of historical progress, have described eighteenth-century mechanical practitioners to be entrepreneurial individuals, or agents of change, who were at odds with the expectations of their time. Many Swedish historians, as well as historians of science and technology, have understood the work of mechanical practitioners through abstractions such as "industrialisation" or "rationalisation". Although such broad generalisations certainly have undeniable appeal, and often a prominent place in concluding chapters such as this one, they are also problematic. When investigating early modern mechanics through the lens of such abstractions, we become blind to the actors' own understanding of what they were doing. First, these abstractions obscure the aspects of early modern mechanics that do not easily fit into their long narratives. When studying early modern mechanics in order to understand the spread of industrialism, it is convenient to ignore how historical actors discussed it as a transformative exercise of virtue, or how they integrated such work into a theoretical order. By ignoring instead the long narrative of industrialism, I have been able to turn my focus to these often-neglected aspects of early modern mechanics. Second, long narratives tend to explain the failure of historical actors teleologically, by relating it to what would come. Such narratives might, for example, explain the failure of mechanical practitioners to deliver on their dreams by showing why certain eighteenth-century societies were not mature enough for modern technology, or that "science and technology" had not developed to the point where they could truly benefit each other. But such an answer seems to beg yet another question: if mechanics was not mature enough to deliver, then why did eighteenth-century states repeatedly reward grand mechanical visions that were hard to realise?

By having focused on the performances of mechanical practitioners in relation to the expectations of contemporaries, I have presented an alternative narrative, which nuances our understanding of early modern mechanics. In the aftermath of the cases presented in this thesis, it is certainly still possible to see the actions of mechanical practitioners, such as Duhre and Polhammar, as part of the processes of modernity. However, it is evident that these actors themselves did not relate their work to the abstractions of modernisation theorists. In order to integrate their own understanding of mechanics into a narrative of industrial progress, we would therefore have to assign them an ironic, or tragic, role. We would have to view them as guardians of an early modern order who were also, nonetheless, instigators of modernity. In other words, if men such as Duhre and Polhammar brought about a modern industrial world, they did so unintentionally while working hard to perfect the status quo of an early modern theocracy. Alternatively, we might write a more straightforward history, which makes these men actors in a history that is more chaotic and non-linear than long narratives can account for.

If I were forced to make a choice between these two general historical frameworks, I would put my faith in the second one. Accordingly, I do not think that the general developments, which I have identified, should necessarily be linked to processes of modernity, ranging over centuries. Such general long-term narratives of modernity provide little understanding of the historical specificity of the changing roles of mathematics in the Swedish state. In order to better understand the ways in which mathematics fostered useful subjects, it would be necessary to study a broader set of institutions in which mathematics and mechanics were part of the education of officials: for example, the fortification corps, the artillery, the navy and the Bureau of Commerce. However, it would not be a stretch to argue that the new role of mathematics in the Swedish state administration of the 1720s, as an exercise in discernment and consent, might be linked to the changing status of the Swedish realm: from a Baltic empire to a post-imperial entity struggling to find a new and stable identity.

This thesis has provided a synchronic understanding of eighteenth-century mechanical practice. I have shown how early moderns understood mechanics not only as a means of making machines, but also as a way to craft pious subjects aligned to a theocratic political order. By ignoring long-term historical change, I have been able to focus on the men, relationships and performances that were forged through mechanical work. Mechanical exercises, and the relationships they established, were believed to foster a pious man: the mechanicus. This mechanical man became a means of imagining a structure where disordered men, a chaotic nature or an unbalanced society were made orderly. If the mechanicus were a man ahead of his time, he belonged to a well-ordered future that would never come to pass.

Glossary

Auskultant	Auscultator
Bergskollegium	Bureau of Mines
Bergsman	Approximately "A man experienced in mining and smelting" (see page 94, note 44)
Bergsråd	Mine Councillor
Bergssciencer	Mining sciences
Bergsväsendet	Mining and smelting
Bergswettenskaper	Mining sciences, or knowledge of mining (depending on context)
Bruk	Ironworks
Brukspatron	Iron master (lit. a patron of a *bruk*)
Cammar- och œconomiedeputationen	Deputation of Cameral Matters and Œconomy
Domprost	Cathedral dean
Fältkansli	Field office
Generalbergsamt	General mining board
Hovrätt	Court of Appeal
Husbonde	Master of the household
Hushåll	House
Häradsrätt	Hundred court
Ingenium / snille	Aptitude
Kammarkollegium	Bureau of Accounts
Kammer	Fiscal chamber
Kanslikollegium	Chancery
Kansliråd	Chancery Councillor
Kollegium	Bureau
Konsistorium	Academic senate
Kungliga vetenskapsakademien	Royal Swedish Academy of Sciences
Kungsladugård	Royal estate
Landshövding	County governor
Länsman	County sheriff
Memorial	Proposal
Pliktpall	Kneeler
Statskontoret	State Office
Wet(t)enskap	Science/knowledge (depending on context, see page 37)

List of illustrations

Cover art. Detail from the frontispiece of Georg Andreas Böckler's *Theatrum Machinarum Novum* (1661). (Photo: Uppsala University Library.)

Figure 1 (page 60). The frontispiece of Anders Celsius' *Arithmetica.* The engraving was signed as invented and made [*inven. et fecit*] by J[an] Klopper, the teacher of drawing at Uppsala, and sculpted [*Sc.*] by Carl Meurman, auscultator in the Bureau of Mines. (Photo: Uppsala University Library.)

Figure 2 (page 90). Proportions of various topics mentioned in the applications to the Bureau of Mines between 1720 and 1750.

Figure 3 (page 90). Table of topics in auscultation applications 1700–50, grouped by decade; relative (and absolute) values.

Figure 4 (page 99). The frontispiece of Fredrik Palmqvist's *Introduction to algebra* (1748). The text reads "In the service of the fatherland." Framed by a laurel are five mathematical works in Swedish, many of them discussed in Chapter 2. Besides Palmqvist's *Introduction to algebra* itself, the books were Celsius' *Arithmentica* (1727), Elvius's treatise on waterwheels *Mathematiskt tractat om effecter af vatn-drifter* (1742), Strömer's translation of Euclid's *Elementa* (1744), and Palmqvist's own treatise on *The solidity and strength of bodies* (1744). (Photo: Uppsala University Library.)

Figure 5, 6 & 7 (page 117). llustrations of mechanical powers, from the margins of Brandt's handwritten manuscript (UUB A 28). From top to bottom: the pulley, the inclined plane (in the form of a wedge) and the screw (Photo: Uppsala University Library.)

Figure 8 (page 131). The cameralist order of society represented in the form of a portable machine model, offered to King Frederik IV of Denmark and Norway around 1700. (Photo: Tor Aas Haug/Norsk Bergverksmuseum.)

Figure 9 (page 133). Engraving enclosed in a letter from Birger Elfwing to Karl XII. (Reproduced from Ernst E. Areen: "Birger Elfving", 19.)

Bibliography

Manuscript sources

Library of the Diocese of Linköping (Linköpings stiftsbibliotek)

Codex Br. 31. Letter from *Collegium Curiosorum* to Casten Feif: "Copy of letter of recommendation for Polhammar", 1711-10-05.

RA: National Archives, Stockholm (Riksarkivet)

Bergskollegium, huvudarkivet

A1. Protocols of the Bureau of Mines
    38–106 (1700–50). Discussions on auscultation and the *Stipendium Mechanicum* etc.
D6. Notes and copies by Daniel Tilas
    14. "Om nuvarande Bergskollegium, dess historia, organisation, verksamhet m.m.", 225–305.
E1. Incoming royal letters
    7 (1698–1700) "Decision on instating a stipendio mechanico", 1699-07-22 [page 594].
E4. Incoming letters, supplements and inquiries
    105–223 (1700–50). Applications to auscultate, applications for the *Stipendium Mechanicum*, requests for study tours, letters from Anders Gabriel Duhre and Carl Duhre.

Frihetstidens utskottshandlingar

Duhre, Anders Gabriel: "Handlingar ang. Ultuna. Ingivna av Anders Gabriel Duhre", 1738–9, R 2702.
————: "Memorial angående Ultuna med bilagor", 1731, R 2556.

Kollegiers m fl, landshövdingars, hovrätters och konsistoriers skrivelser till Kungl Maj:t.

8. Letters from the Bureau of Mines to the king
    7. "On stipends to the auscultators", 1691-06-16.
    19. (1719–22) "Concerning rank", 1719-06-01.
    ———— "On Carl Duhre", 1720-01-19.
    ———— "On Carl Duhre, continued", 1721-04-17.
    22 (1730–4). "Concerning rank", 1730-01-26.

ULA: Regional Archives of Uppsala (Landsarkivet i Uppsala)

Ulleråkers häradsrätt
A1:2. "Dombok från Ulleråker ting (1711–29).

KB: Royal Library, Stockholm (Kungliga biblioteket)
Karl XII: "Copy of letter to the council concerning new privileges for Stiernsund",
I p:23 1, 1712-03-13 [page 59].
Bureau of Mines: "Letter to Christopher Polhem", 1716-04-23, I p:23 1 [page 101].
Casten Feif: "Letter to Christopher Polhem", 1714-12-20, I p:23 1 [page 86].
Nils Stromberg: "Letter to Christopher Polhem concerning the King's recommendation
of Polhammar", 1712-06-03, I p:23 1 [page 64].
———: "Letter to Christopher Polhem", 1714-10-06, I p:23 1 [page 82].

Uppsala University Library (UUB) & Uppsala University Archives
(UUA)

Collegiernas bref & resolutioner för Uppsala Universitet 1623–1733, UUA
Letter from Kammarkollegium to the academic senate of Uppsala University,
1725-09-04.

Handwritten manuscripts
Brandt, Georg: "Kort handledning till Driftkonsten eller Mechaniken", A 28 [page 245–
302].
———: "Om geometrien", A 28 [page 41–81].
———: "Om perspektivkonsten", A 28 [page 538–67].
———: "[On Philosophia naturali]", A 28 [page 334–465].
Duhre, Anders Gabriel: "Förtahl. Hwilket And. Gab. Duhre är sinnad att sättia
framman för sitt mathematiska wärk, som han nu hafwer under händer uti upsååt
at låta och uthgå på wårt Swenska Moders mål", A 29 [≈1714].
Polhem, Christopher: "Betyg för Andreas Duhre", 1712–07–30, X 241:Duhre.
Lars Schultze: "Kort betänkande huru en yngling som tänker söka sin fortkomst vid
bergsväsendet bör sin tid anlägga", D 1433, UUB.
———: "Kort handledning till Driftkonsten eller Mechaniken", D 1433, UUB.
———: "Om geometrien", C 1433, UUB.

Online resources

SAOB: Svenska Akademins ordbok, http://g3.spraakdata.gu.se/saob/
National Museum of Science and Technology, Stockholm: "Christopher Polhem.
Tillbaka till framtiden" <http://www.tekniskamuseet.se/1/2604.html> [accessed
27 March 2015].

Printed sources and literature

Acton, Alfred (ed): *The letters and memorials of Emanuel Swedenborg* vol. 1, 1709–48 (Bryn Athyn PA., 1948).

Agner, Eric: *Arithmetica fractionum. Thet är: räkne-konst vthi brutne-tahl* (Stockholm, 1710).

————: *Geodæsia Suecana eller Örtuga delo-bok* (Stockholm, 1730).

Ahlström, Walter: *Arvid Horn och Karl XII 1710–1713* (Lund, 1959).

Alder, Ken: *Engineering the revolution. Arms and Enlightenment in France 1763–1815* (Princeton NJ, 1997).

Algazi, Gadi: "Scholars in households. Refiguring the learned habitus, 1480–1550", *Science in context* 16:1–2 (2003), 9–42.

Almquist, Johan Axel: *Bergskollegium och Bergslagsstaterna 1637–1857. Administrativa och biografiska anteckningar* (Stockholm, 1909).

Almquist, Karl Gustaf: *Andreas Rydelius' etiska åskådning* (Lund, 1955).

Andersson, Gustaf & Carl Gustaf von Brinkman (eds): *Handlingar ur v. Brinkmanska archivet på Trolle-Ljungby* vol. 1 (Örebro, 1859).

Alströmer, Jonas: *Sveriges wälstånd om det will* (Stockholm, 1745).

Althin, Torsten: *Christopher Polhem och Stjernsunds manufacturverk* (Säter, 1950).

Amelin, Olov: *Medaljens baksida. Instrumentmakaren Daniel Ekström och hans efterföljare i 1700-talets Sverige* (Uppsala, 1999).

Andersson, Gustaf, and Carl Gustaf von Brinkman (eds): *Handlingar ur v. Brinkmanska archivet på Trolle-Ljungby* (Örebro, 1859).

Annerstedt, Claes: *Upsala universitets historia* 3 vols. (Uppsala, 1908).

Anonymous: *In Andreæ Gabrielis Duhre opusculum svethicum, dictum Summa Sveciæ felicitas, Svethicè, Sweriges högsta wälstånd* (Stockholm, 1738).

————: *In opus Andreæ Gabrielis Duhre, summa felicitas Sveciæ, dictum, Svetice, Sueriges högsta wälstånd* (No place, undated).

————: *Samtal emellan en herre och en fru, om geometriens nytta för unga studerande* (Stockholm, 1743).

————: *Breve ragguaglio della vita e della morte del conte Giovanni Patkul, nobile di Livonia, tradotto dal idioma inglese per illustrazione della vita di Pietro il Grande … e di Carlo duodecimo* (Lugano, 1761).

Apostolidès, Jean-Marie: *Le roi-machine. Spectacle et politique au temps de Louis XIV* (Paris, 1981).

Areen, Ernst E.: "Birger Elfving. Hedemora gevärsfaktori och Furudals styckebruk", *Svenska vapenhistoriska sällskapets årsskrift* 3 (1931), 3–76.

Ashworth, William J.: "The ghost of Rostow. Science, culture and the British industrial revolution", *History of science* 46:3 (2008), 249–74.

————: "The British industrial revolution and the ideological revolution. Science, neoliberalism and history", *History of science* 52:2 (2014), 178–99.

Aurelius, Aegidius Matthiæ: *Arithmetica eller een kort och eenfaldigh räknebook, uthi heele och brutne taal. Medh lustige och sköne exempel, them eenfaldighom som til thenne konst lust och behagh hafwe* (Uppsala, 1614).

Barker, Peter, and Bernard R. Goldstein: "Theological foundations of Kepler's astronomy", *Osiris* 16 (2001), 88–113.

Barnes, Barry: "On the conventional character of knowledge and cognition", *Philosophy of the social sciences* 11:3 (1981), 303–33.

Beik, William: "The absolutism of Louis XIV as social collaboration", *Past & Present* 188:1 (2005), 195–224.

Bennett, Jim: "The mechanical arts", in Katharine Park and Lorraine Daston (eds): *The Cambridge history of science* vol. 3: early modern science (Cambridge, 2008), 673–95.

———: "The mechanics' philosophy and the mechanical philosophy", *History of science* 24 (1986), 1–28.

Berch, Anders: *Åminnelse-tal, öfver commerce-rådet Henric Kalmeter, efter kongl. vetensk. academiens befallning, hållit i stora riddarhus-salen den 13. martii, 1752* (Stockholm, 1752).

Bergman, Torbern: *Åminnelse-tal öfver framledne bergs-rådet och medicinæ doctoren … Georg Brandt, hållet för kongl. vetenskaps academien, uti stora riddarehus-salen, den 9 aug. 1769* (Stockholm, 1769).

Biagioli, Mario: *Galileo, courtier. The practice of science in the culture of absolutism* (Chicago, 1993).

———: "The social status of Italian mathematicians, 1450–1600", *History of Science* 27:1 (1989), 41–95.

Bijker, Wiebe, and Trevor Pinch: "The social construction of facts and artifacts. Or how the sociology of science and the sociology of technology might benefit each other", in Wiebe Bijker, Thomas Hughes, and Trevor Pinch (eds): *The social construction of technological systems. New directions in the sociology and history of technology* (Cambridge MA, 1987), 17–50.

Bloor, David: "Wittgenstein and Mannheim on the sociology of mathematics", *Studies in history and philosophy of science. Part A* 4:2 (1973), 173–91.

Boethius, B: "Aegidius Matthie Upsaliensis Aurelius", *SBL* 2 (Stockholm, 1920), 451–4.

Bohnstedt, M: "Från Bendertiden. Casten Feifs brev till Arvid Horn 1710–12", *Personhistorisk tidskrift* (1921).

Bourdieu, Pierre, Olivier Christin, and Pierre-Etienne Will: "Sur la science de l'État", *Actes de la recherche en sciences sociales* 133:1 (2000), 3–11.

Brandt, Georg: *En grundelig anledning til mathesin universalem och algebram, efter herr And. Gabr. Duhres håldne prælectioner sammanskrifwen* (Stockholm, 1718).

Brianta, Donata: "Education and training in the mining industry, 1750–1860. European models and the Italian case", *Annals of science* 57:3 (2000), 267–300.

Bring, Samuel E.: "Bidrag till Christopher Polhems lefnadsteckning", *Christopher Polhem. Minnesskrift utgifven av Svenska teknologföreningen* (Stockholm, 1911).

———: *Christopher Polhem. The father of Swedish technology* (Hartford CT, 1963).

———: "Några bref från Casten Feif till Christopher Polhem", *Karolinska förbundets årsbok* (1911), 233–56.

Brobergs, Gunnar: "Then gifte Philosophen eller en man som blyges att wara gift", *Annales Academiæ regiæ scientiarum Upsaliensis:s. Kongl. Vetenskapssamhällets i Uppsala Årsbok* 26 (1985–86), 72–91.

Brunius, Teddy: *Andreas Rydelius och hans filosofi* (Lund, 1958).

214

Burke, Peter: *The fabrication of Louis XIV* (New Haven, 1992).

Butler, Judith: "Performative acts and gender constitution. An essay in phenomenology and feminist theory", in Henry Bial (ed): *The performance studies reader*, 2nd edition (London & New York, 2007).

Carlqvist, Gunnar: "Karl XII:s ungdom och första regeringsår", in Samuel E. Bring (ed): *Karl XII. Till 200-årsdagen av hans död* (Stockholm, 1918), 43–86.

Carter, Philip: *Men and the emergence of polite society. Britain 1660–1800* (Harlow, 2001).

————: "James Boswell's manliness", in Tim Hitchcock and Michèle Cohen (eds): *English masculinities 1660–1800* (London & New York, 1999), 111–130

Cavallin, Maria: *I kungens och folkets tjänst. Synen på den svenske ämbetsmannen 1750–1780* (Gothenburg, 2003).

Celsius, Anders: *Arithmetica eller Räkne-konst, grundeligen demonstrerad af Anders Celsius* (Uppsala, 1727).

Celsius, Olof: *Åtminnelse-tal öfver lands-höfdingen, bärgs-rådet Lars Benzelstierna, hållit för Kongl. vetenskapsacademien den 6 dec. 1758* (Stockholm, 1759).

de Certeau, Michel: *The practice of everyday life* (Berkeley & Los Angeles, 1984).

Chang, Hasok: "We have never been whiggish (about phlogiston)", *Centaurus* 51:4 (2009), 239–64.

Christensen, Dan Ch.: *Det moderne projekt. Teknik & kultur i Danmark-Norge 1750– (1814)–1850* (Copenhagen, 1996).

Clairaut, Alexis Claude and Pehr Elvius (tran.): *Inledning til geometrien. Af herr Clairaut* (Stockholm, 1744).

Clark, William: *Academic charisma and the origins of the research university* (Chicago, 2006).

Cohen, Michèle: "'Manners' make the man. Politeness, chivalry, and the construction of masculinity, 1750–1830", *Journal of British studies* 44:2 (2005), 312–29.

Collins, Harry M.: *Changing order. Replication and induction in scientific practice* (Chicago, 1992).

————: "Learning through enculturation", in Angus Gellatly, Don Rogers, and John A. Sloboda (eds): *Cognition and social worlds* (Oxford, 1989), 205–15.

Collins, H. M.: "The seven sexes. A study in the sociology of a phenomenon, or the replication of experiments in physics", *Sociology* 9:2 (1975), 205–24.

Condren, Conal: "The persona of the philosopher and the rhetorics of office in early modern Europe", in Conal Condren, Stephen Gaukroger, and Ian Hunter (eds): *The philosopher in early modern Europe. The nature of a contested identity* (Cambridge, 2006), 66–89.

Condren, Conal, Stephen Gaukroger, and Ian Hunter: "Introduction", in Conal Condren, Stephen Gaukroger, and Ian Hunter (eds): *The philosopher in early modern Europe. The nature of a contested identity* (Cambridge, 2006), 1–16.

Connell, Raewyn: *Masculinities* (Cambridge, 2005).

Cooper, Alix: *Inventing the indigenous. Local knowledge and natural history in early modern Europe* (Cambridge, 2007).

————: "'The possibilities of the land'. The inventory of 'natural riches' in the early modern German territories", in Margaret Schabas and Neil De Marchi (eds): *Oeconomies in the age of Newton* (Durham NC, 2004), 129–53.

Corneanu, Sorana: *Regimens of the mind. Boyle, Locke, and the early modern cultura animi tradition* (Chicago, 2011).

Cronstedt, Axel Fredrik: *Åminnelse-tal öfver framledne directeuren och kongl. vetensk. acad. ledamot Henric Theoph. Scheffer, på kongl. vetensk. acad. vägnar, hållit i stora riddarehus-salen, den 17. september 1760* (Stockholm, 1760).

Dahlgren, Stellan: "Karl XI:s envälde - kameralistisk absolutism?", *Makt & vardag* (Stockholm, 1993), 115–32.

Dahl, Per: *Svensk ingenjörskonst under stormaktstiden. Olof Rudbecks tekniska undervisning och praktiska verksamhet* (Uppsala, 1995).

Dalin, Olof von: *Åminnelse-tal öfver kongl. vetenskaps academiens medlem och secreterare, herr Pehr Elvius* (Stockholm, 1750).

———: *Then swänska Argus* no. 5 (Stockholm, 1732).

Danielsson Malmros, Ingmarie: *Det var en gång ett land. Berättelsen om svenskhet i historieläroböcker och elevers föreställningsvärldar* (Lund, 2012).

Daston, Lorraine: "Enlightenment calculations", *Critical inquiry* 21:1 (1994), 182–202.

Daston, Lorraine, and Peter Gailson: *Objectivity* (New York, 2007).

Daston, Lorraine, and Otto Sibum: "Introduction. Scientific personae and their histories", *Science in context* 16:1 (2003), 1–8.

Dear, Peter: *Discipline & experience. The mathematical way in the scientific revolution* (Chicago, 1995).

———: "Mysteries of state, mysteries of nature. Authority, knowledge and expertise in the 17th Century", in Sheila Jasanoff (ed): *States of knowledge. The co-production of science and social order* (London, 2004), 206–24.

———: *The intelligibility of nature. How science makes sense of the world* (Chicago, 2006).

Deason, Gary B.: "Reformation theology and the mechanistic conception of nature", in David C. Lindberg and Ronald L. Numbers (eds): *God and nature. Historical essays on the encounter between Christianity and science* (Berkeley, 1986).

Duhre, Anders Gabriel: *And. Gab. Duhres Ödmiukaste memoriale stäldt til Swea rikes samtel. högloflige ständer wid 1726 åhrs riks-dag* (Stockholm, 1726).

———: *Förklaring öfver des tilförende uthgifne Wälmente tanckar angående huru han tillika med sin broder är sinnad at utan almenna bestas betungande uppå deras egit äfwentyr uprätta ett laboratorium mathematico-oeconomicum* (Stockholm, 1722).

———: *Första delen af en grundad geometria, bewijst uti de föreläsningar, som äro håldne på swänska språket uppå kongl. fortifications contoiret i Stockholm* (Stockholm, 1721).

———: *Sweriges högsta wälstånd, bygdt uppå en oeconomisk grundwal* (Stockholm, 1738).

———: *Tanckar, angående huruledes man i mangel af strömar och fall måtte med en synnerlig förmån, allestädes (hwarest stilla stående watn finnes) kunna drifwa allehanda rörliga wärck* (Stockholm, 1723).

———: *Utdrag af Anders Gabriel Duhres oeconomiske tractat, kallad Sweriges högsta wälstånd, hwarutinnan sammanhanget så wäl som sakens höga wigt och angelägenhet, til alla wälsinte patrioters widare ompröfwande, warder kort och tydeligen å:daga lagd* (Stockholm, 1738).

———: *Wälmenta tanckar, angående huru jag tillika med min broder är sinnad, at utan almenna bestas betungande, uppå wårt egit äfwentyr uprätta et laboratorium mathematico-oeconomicum* (Stockholm, 1722).

Dunér, David: "Sextiofyra och åtta istället för tio. Karl XII, Swedenborg och konsten att räkna", *Scandia* 67 (2001), 211–38.

⸻: *Tankemaskinen. Polhems huvudvärk och andra studier i tänkandets historia* (Nora, 2012).

⸻: *The natural philosophy of Emanuel Swedenborg. A study in the conceptual metaphors of the mechanistic world-view* (Dordrecht, 2013).

⸻: *Världsmaskinen. Emanuel Swedenborgs naturfilosofi* (Lund, 2004).

Edgerton, David: "Innovation, technology, or history. What is the historiography of technology about?", *Technology and culture* 51:3 (2010), 680–97.

⸻: *The shock of the old. Technology and global history since 1900* (New York, 2007).

Eisenstadt, S. N., and Louis Roniger: "Patron–client relations as a model of structuring social exchange", *Comparative studies in society and history* 22:1 (1980), 42–77.

Ekedahl, Nils: *Det svenska Israel. Myt och retorik i Haquin Spegels predikokonst* (Uppsala, 1999).

Eklund, Eric: *Upfostrings-läran, som wisar sätt och medel til ungdomens rätta skiötsel och underwisning* (Stockholm, 1746).

Ekström, Daniel: *Tal, om järn-förädlingens nytta och vårdande* (Stockholm, 1750).

Ellenius, Allan: *Karolinska bildidéer* (Uppsala, 1966).

Elvius, Pehr: *Mathematisk tractat, om effecter af vatn-drifter, efter brukliga vatn-värks art och lag* (Stockholm, 1742).

Euklides, and Mårten Strömer (tran.): *Euclidis Elementa eller grundeliga inledning til geometrien, til riksens ungdoms tienst på svenska språket utgifwen* (Uppsala, 1744).

Fletcher, Anthony: *Gender, sex and subordination in England 1500–1800* (New Haven, 1995).

⸻: *Growing up in England. The experience of childhood, 1600–1914* (New Haven, 2008).

Florén, Anders, and Göran Rydén: *Arbete, hushåll och region. Tankar om industrialiseringsprocesser och den svenska järnhanteringen* (Uppsala, 1992).

Fors, Hjalmar: "J. G. Wallerius and the laboratory of enlightenment", in Hjalmar Fors, Enrico Baraldi, and Anders Houltz (eds): *Taking place. The spatial contexts of science, technology and business* (Sagamore Beach MA, 2006), 3–33.

⸻: "Kemi, paracelsism och mekanisk filosofi. Bergskollegium och Uppsala cirka 1680–1770", *Lychnos* (2007), 165–98.

⸻: *Mutual favours. The social and scientific practice of eighteenth-century Swedish chemistry* (Uppsala, 2003).

⸻: *The limits of matter. Chemistry, mining, and Enlightenment* (Chicago, 2015).

Frängsmyr, Tore: *Wolffianismens genombrott i Uppsala. Frihetstida universitetsfilosofi till 1700-talets mitt* (Uppsala, 1972).

Friedrich II of Prussia: "Considérations sur l'état présent du corps politique de l'Europe", *Oeuvres posthumes de Frédric II, roi de Prusse* vol. 6 (Berlin, 1788 [1736]), 3–52.

Fuchs, Thomas: *The mechanization of the heart. Harvey and Descartes* (Rochester, N.Y, 2001).

Gabbey, Alan: "Between ars and philosophia naturalis. Reflections on the historiography of early modern mechanics", in J. V. Field and Frank A. J. L. James (eds): *Renaissance and revolution. Humanists, scholars, craftsmen, and natural philosophers in early modern Europe* (Cambridge, 1993), 133–45.

Garber, Daniel: "Remarks on the pre-history of the mechanical philosophy", in Sophie Roux (ed): *The mechanization of natural philosophy* (Dordrecht, 2013), 3–26.

Gaukroger, Stephen: *Francis Bacon and the transformation of early-modern philosophy* (Cambridge, 2001).

————: "The persona of the natural philosopher", in Conal Condren, Stephen Gaukroger, and Ian Hunter (eds): *The philosopher in early modern Europe. The nature of a contested identity* (Cambridge, 2006), 17–34.

Gaunt, David: *Utbildning till statens tjänst. En kollektivbiografi av stormaktstidens hovrättsauskultanter* (Uppsala, 1975).

Gauvin, Jean-François: *Habits of knowledge. Artisans, savants and mechanical devices in seventeenth-century French natural philosophy* (Cambridge MA, 2008).

Genette, Gerard: *Paratexts. Thresholds of interpretation* (Cambridge, 1997).

Gillispie, Charles Coulston: *Science and polity in France at the end of the old regime* (Princeton NJ, 1980).

Goffman, Erving: *The presentation of self in everyday life* (London, 1990).

Golinski, Jan: *Science as public culture. Chemistry and Enlightenment in Britain, 1760–1820* (Cambridge, 1992).

————: "The care of the self and the masculine birth of science", *History of science* 40 (2002), 125–45.

Goodman, Dena: *The republic of letters. A cultural history of the French Enlightenment* (Ithaca, 1994).

Grape, Anders: *Ihreska handskriftssamlingen i Uppsala universitets bibliotek*, vol. 2 (Uppsala, 1949).

————: *Något om Anders Gabriel Duhre och en honom ägnad latinsk dikt* (Stockholm, 1949).

Grauers, Sven: "Karl XII:s personlighet. Försök till analys", *Karolinska förbundets årsbok* (1969), 7–82.

Greenblatt, Stephen: *Renaissance self-fashioning. From More to Shakespeare* (Chicago, 2005 [1980]).

Hackett, Jeremiah: "Roger Bacon on scientia experimentalis", in Jeremiah Hackett (ed): *Roger Bacon and the sciences. Commemorative essays* (Leiden, 1997), 278–316.

Hacking, Ian: "Artificial phenomena. Essay review of *Leviathan and the air pump*", *The British journal for the history of science* 24:02 (1991), 235–41.

Hagen, Lorentz: *Anecdotes concerning the famous John Reinhold Patkul. Or, an authentic relation of what passed betwixt him and his confessor, the night before and at his execution. Translated from the original manuscript, never yet printed* (London, 1761).

————: *Das schemertzliche doch seelige Ende, des welt-bekandten Joh. Reinhold Patkuls* (Cologne, 1714).

————: *Een kort verhaal, wegens de verschrikkelyke dood, van den vermaarden heer Johann Reinold van Patkul, generaal van zyn koninklyke majesteyt van Zweeden* (Amsterdam, 1718).

Hamrén, Olof: *Manufacturs-spegel, som genom tilförlåtelige uträkningar och uprichtiga grundsatzer gifwer et fulkommeligit lius om manufacturers och fabriquers rätta art och beskaffenhet, samt oskattbara nytta genom försiktig anlägning* (Stockholm, 1738).

Hanley, Sarah: "Engendering the state. Family formation and state building in early modern France", *French historical studies* 16:1 (1989), 4–27.

Hård, Mikael: "Mechanica och mathesis. Några tankar kring Christopher Polhems fysikaliska och vetenskapsteoretiska föreställningar", *Lychnos* (1986), 55–69.

Harris, Frances: "Living in the neighbourhood of science. Mary Evelyn, Margaret Cavendish and the Greshamites", in Lynette Hunter and Sarah Hutton (eds): *Women, science and medicine 1500–1700. Mothers and sisters of the Royal Society* (Stroud, 1997), 198–217.

Hatton, Ragnhild M.: *Charles XII of Sweden* (London, 1968).

Hebbe, Per Magnus: "Anders Gabriel Duhres 'Laboratorium mathematico-oeconomicum'. Ett bidrag till Ultunas äldre historia", *Kungl. Landtbruks-akademiens handlingar och tidskrift* 72 (1933), 576–94.

Hedström, Jan-Olof: *-igenom gode ordningar och flitigt upseende-. Bergsstaten 375 år* (Uppsala, 2012).

Heilbron, John Lewis: *Geometry civilized. History, culture, and technique* (Oxford, 2000).

Helander, Hans: "Introduction", in Emanuel Swedenborg: *Festivus applausus in Caroli XII* (Uppsala, 1985), 9–50.

Hessenbruch, Arne: "The spread of precision measurement in Scandinavia 1660–1800", in Kostas Gavroglu (ed): *The sciences in the European periphery during the Enlightenment* (Dordrecht, 1999), 179–224.

Heyman, H. J.: "Anders Celsius", *SBL* 8 (Stockholm, 1929), 266.

Hilaire-Pérez, Liliane: "Dissemination of technical knowledge in the Middle Ages and the early modern era. New approaches and methodological issues", *Technology and culture* 47:3 (2006), 536.

————: *L'invention technique au siècle des Lumières* (Paris, 2000).

————: "Technology as a public culture in the eighteenth century. The artisan's legacy", *History of science* 45 (2007), 135–53.

Hildebrand, Bengt: "Anders Gabriel Duhre", *SBL* 11 (Stockholm, 1945), 506.

————: "Casten Feif", *SBL* 15 (Stockholm, 1956), 512.

————: "Duhre, släkt", *SBL* 11 (Stockholm, 1945), 505.

————: *Kungl. Svenska Vetenskapsakademien. Förhistoria, grundläggning och första organisation* vol. 1 (Stockholm, 1939).

Hildebrand, Ingegerd: "Falkenberg Af Sandemar, Gabriel", *SBL* 15 (Stockholm, 1956), 222–7.

Hill, Katherine: "'Juglers or schollers?'. Negotiating the role of a mathematical practitioner", *The British journal for the history of science* 31:03 (1998), 253–74.

Hinners, Linda: *De fransöske handtwerkarne vid Stockholms slott 1693–1713. Yrkesroller, organisation, arbetsprocesser* (Stockholm, 2012).

Hodacs, Hanna, and Kenneth Nyberg: *Naturalhistoria på resande fot. Om att forska, undervisa och göra karriär i 1700-talets Sverige* (Stockholm, 2007).

Högberg, Staffan: "Inledning", *Anton von Swabs berättelse om Avesta kronobruk 1723* (Stockholm, 1983), 7–30.

Höpken, Anders Johan von: *Åminnelse-tal öfver astronomiæ professoren Anders Celsius efter kongl. vetenskaps academiens befalning hållit i stora riddarhus-salen d. 27 novemb. 1745.* (Stockholm, 1745).

Ihalainen, Pasi: "New visions for the future. Bodily and mechanical conceptions of the political community in eighteenth-century Sweden", in Petri Karonen (ed): *Hopes and fears for the future in early modern Sweden, 1500–1800* (Helsinki, 2009), 315–40.

Ingemarsdotter, Jenny: *Ramism, rhetoric and reform. An intellectual biography of Johan Skytte (1577–1645)* (Uppsala, 2011).

Israel, Jonathan I.: *Radical enlightenment. Philosophy and the making of modernity 1650–1750* (Oxford, 2001).

Jacob, Margaret C.: *The first knowledge economy. Human capital and the European economy, 1750–1850* (Cambridge, 2014).

Jacob, Margaret C., and Larry Stewart: *Practical matter. Newton's science in the service of industry and empire, 1687–1851* (Cambridge MA, 2004).

Jansson, Karin Hassan: "Ära och oro. Sexuella närmanden och föräktenskapliga relationer i 1700-talets Sverige", *Scandia* 75:1 (2009), 29–56.

————: *Kvinnofrid. Synen på våldtäkt och konstruktionen av kön i Sverige 1600–1800* (Uppsala, 2002).

Johannesson, Kurt: *I polstjärnans tecken. Studier i svensk barock* (Stockholm, 1968).

————: "Om furstars och aristokraters dygder. Reflexioner kring Johannes Schefferus Memorabilia", in Sten Åke Nilsson and Margareta Ramsay (eds): *1600-talets ansikte* (Nyhamnsläge, 1997), 309–21.

Johannisson, Karin: "Naturvetenskap på reträtt. En diskussion om naturvetenskapens status under svenskt 1700-tal", *Lychnos* (1979), 107–54.

Johns, Adrian: "The ambivalence of authorship in early modern natural philosophy", in Mario Biagioli and Peter Galison (eds): *Scientific authorship. Credit and intellectual property in science* (New York, 2003), 67–90.

————: *The nature of the book. Print and knowledge in the making* (Chicago, 1998).

Johnston, Stephen: "Mathematical practitioners and instruments in Elizabethan England", *Annals of science* 48:4 (1991), 319–44.

Jones, Matthew L.: *The good life in the scientific revolution. Descartes, Pascal, Leibniz, and the cultivation of virtue* (Chicago, 2006).

Josephson, Ragnar: "Karl XI och Karl XII som esteter", *Karolinska förbundets årsbok* (1947), 7–67.

Justi, Johann Heinrich Gottlob von: *Der Grundriss einer guten Regierung in fünf Buchern* (Frankfurt and Leipzig, 1759).

Kaiserfeld, Thomas: *Krigets salt. Salpetersjudning som politik och vetenskap i den svenska skattemilitära staten under frihetstid och gustaviansk tid* (Lund, 2009).

Kantorowicz, Ernst Hartwig: *The king's two bodies. A study in mediaeval political theology* (Princeton NJ, 1957).

Karlsson, Åsa: *Den jämlike undersåten. Karl XII:s förmögenhetsbeskattning 1713* (Uppsala, 1994).

————: "En man i statens tjänst. Den politiska elitens manlighetsideal under det karolinska enväldet 1680–1718", in Anne Marie Berggren (ed): *Manligt och omanligt i ett historiskt perspektiv* (Uppsala, 1999), 116–28.

Klingenstierna, Samuel: *Åminnelse-tal öfver kongl. vetensk. academiens framledne ledamot, commerce-rådet Christopher Polhem, på kongl. vetenskaps academiens vägnar hållit i stora riddarhus-salen* (Stockholm, 1753).

Koepp, Cynthia J.: "The alphabetical order. Work in Diderot's Encyclopédie", in Steven L. Kaplan and Cynthia J. Koepp (eds): *Work in France. Representations, meaning, organization, and practice* (Ithaca, 1986), 229–57.

Koskinen, Ulla: "'Benevolent lord' and 'willing servant'. Argumentation with social ideals in late-sixteenth-century letters", in Petri Karonen (ed): *Hopes and fears for the future in early modern Sweden, 1500–1800* (Helsinki, 2009), 55–76.

La Motraye, Aubry de: *Voyages du sr. A. de La Motraye, en Europe, Asie & Afrique; où l'on trouve une grande variété de recherches geographiques, historiques & politiques* vol. 2 (The Hague, 1727).

Laborier, Pascale: "Les sciences camérales, prolégomènes à toute bureaucratie future ou parades pour gibiers de potence?", in Pascale Laborier, Frédéric Audren, Paolo Napoli, and Jakob Vogel (eds): *Les sciences camérales. Activités pratiques et histoire des dispositifs publics* (Paris, 2011), 11–30.

Laird, Walter Roy: "Robert Grosseteste on the subalternate sciences", *Traditio* 43 (1987), 147–69.

Landahl, Sten: *Bondeståndets riksdagsprotokoll* vol. 1 [1720–7] (Stockholm, 1939).

————: *Bondeståndets riksdagsprotokoll* vol. 2 [1731–34] (Stockholm, 1945).

————: *Sveriges ridderskaps och adels riksdagsprotokoll från och med år 1719* vol. 2 [1723] (Stockholm, 1876).

————: *Sveriges ridderskaps och adels riksdagsprotokoll från och med år 1719* vol. 5 [1726–7] (Stockholm, 1879).

————: *Sveriges ridderskaps och adels riksdagsprotokoll från och med år 1719* vol. 6 [1731] (Stockholm, 1887).

Lappalainen, Mirkka: "Släkt och stånd i Bergskollegium före reduktionstiden", *Historisk tidskrift för Finland* 87:2 (2002), 145–72.

Latour, Bruno: *Science in action. How to follow scientists and engineers through society* (Cambridge MA, 1987).

————: *The pasteurization of France* (Cambridge MA, 1988).

Laurel, Lars: *Åminnelse-tal öfver capitaine mechanichus vid fortificationen … Mårten Trievald, hållet på store riddar-hus salen, den 23 decemb. 1747* (Stockholm, 1748).

Lave, Jean, and Etienne Wenger: *Situated learning. Legitimate peripheral participation* (Cambridge, 2005).

Lawrence, Christopher, and Steven Shapin: "Introduction. The body of knowledge", in Christopher Lawrence and Steven Shapin (eds): *Science incarnate. Historical embodiments of natural knowledge* (Chicago, 1998), 1–19.

————: *Science incarnate. Historical embodiments of natural knowledge* (Chicago, 1998).

Liedman, Sven-Eric: *Den synliga handen. Anders Berch och ekonomiämnena vid 1700-talets svenska universitet* (Stockholm, 1986).

221

Liljecrantz, Axel Johan Carl Vilhelm: "Polhem och grundandet av Sveriges första naturvetenskapliga samfund. Jämte andra anteckningar rörande Collegium Curiosorum. I", *Lychnos* (1939), 289–308.

————: "Polhem och grundandet av Sveriges första naturvetenskapliga samfund. Jämte andra anteckningar rörande Collegium Curiosorum. II", *Lychnos* (1940), 21–54.

————:: *Christopher Polhems brev* (Uppsala, 1941).

Liljegren, Bengt: *Karl XII. En biografi* (Lund, 2000).

————: *Karl XII i Lund. När Sverige styrdes från Skåne* (Lund, 1999).

Lindberg, Bo: *Den antika skevheten. Politiska ord och begrepp i det tidig-moderna Sverige* (Stockholm, 2006).

Lindgren, Michael: "Christopher Polhem", *SBL* 29 (Stockholm, 1995–7), 338.

————: *Christopher Polhems testamente. Berättelsen om ingenjören, entreprenören och pedagogen som ville förändra Sverige* (Stockholm, 2011).

Lindgren, Michael, and Per Sörbom: *Christopher Polhem 1661–1751. "The Swedish Daedalus"* (Stockholm, 1985).

Lindqvist, Janne: *Dygdens förvandlingar. Begreppet dygd i tillfällestryck till handelsmän före 1770* (Uppsala, 2002).

Lindqvist, Svante: *Technology on trial. The introduction of steam power technology into Sweden, 1715–1736* (Uppsala, 1984).

Lindroth, Sten: *Gruvbrytning och kopparhantering vid Stora Kopparberget intill 1800-talets början. 2, Kopparhanteringen* (Uppsala, 1955).

————: *Kungl. Svenska vetenskapsakademiens historia 1739–1818. 1:1, Tiden intill Wargentins död (1783)* (Stockholm, 1967).

————: "Samuel Klingenstierna", *SBL* 21, (Stockholm, 1975–7), 319.

————: *Svensk lärdomshistoria. Frihetstiden* (Stockholm, 1997 [1978]).

Linnaeus, Carl: "Tankar om grunden till œconomien genom naturkunnigheten och physiquen", *Kungliga vetenskapsakademins handlingar* (1739–40), 411–29.

Locke, John and Matthias Riben (tran.): *Herr Johan Lockes tankar och anmärkningar angående ungdomens uppfostring först skrefne uti engelskan, men nu för deras serdeles wärde och nyttighet uppå swenska öfwersatte* (Stockholm, 1709).

Lundgren, Anders: "Gruvor och kemi under 1700-talet i Sverige. Nytta och vetenskap", *Lychnos* (2008), 7–42.

Lundin, Sverker: *Skolans matematik. En kritisk analys av den svenska skolmatematikens förhistoria, uppkomst och utveckling* (Uppsala, 2008).

MacLeod, Christine: "Concepts of invention and the patent controversy in Victorian Britain", in Robert Fox (ed): *Technological change. Methods and themes in the history of technology* (Amsterdam, 1996), 137–53.

————: *Heroes of invention. Technology, liberalism and British identity, 1750–1914* (Cambridge, 2007).

————: *Inventing the industrial revolution. The English patent system, 1660–1800* (Cambridge, 1988).

MacLeod, Christine, and Alessandro Nuvolari: *"The ingenious crowd". A critical prosopography of British inventors, 1650–1850* [working paper] (Eindhoven, 2005).

Magnusson, Lars: *Mercantilism. The shaping of an economic language* (London, 1994).

Mansén, Elisabeth: "Samtal emellan en Herre och en Fru år 1743" in Emma Hagström Molin and Andreas Hellerstedt (eds): *Lärda samtal. En festskrift till Erland Sellberg* (Lund, 2014), 149–163.

Marklund, Andreas: *In the shadows of his house. Masculinity & marriage in Sweden, c. 1760 to the 1830s* (Florence, 2002).

Mauss, Marcel: *Les techniques du corps* (Chicoutimi, 2002 [1934]) <http://classiques.uqac.ca/classiques/mauss_marcel/socio_et_anthropo/6_Tech niques_corps/techniques_corps.pdf> [accessed 27 April 2012].

Mayr, Otto: *Authority, liberty & automatic machinery in early modern Europe* (Baltimore, 1986).

Mennander, Carl Fredrik: *Åminnelse-tal öfver theol. doct. dom-probsten och theol. professor primarius vid kongl. academien i Åbo, herr Gabriel Lauræus, på kongl. vetenskaps academiens vägnar, hållit den 19 december 1755* (Stockholm, 1756).

Merchant, Carolyn: *The death of nature. Women, ecology, and the scientific revolution* (San Francisco, 1980).

Missner, Marshall: "Skepticism and Hobbes's political philosophy", *Journal of the history of ideas* 44:3 (1983), 407–27.

Moberg, Gustaf: *Från exercitier till modern idrott* (Uppsala, 1950).

Mokyr, Joel: *The gifts of Athena. Historical origins of the knowledge economy* (Princeton NJ, 2005).

———: *The lever of riches. Technological creativity and economic progress* (New York, 1990).

Mörk, Jacob Henrik: *Åminnelse-tal öfver framledne theologie doctoren och kyrko-herden i Bolstad i Carlstads stift, herr Nils Brelin, på kongl. vetenskaps academiens vägnar, hållit uti stora riddare- hus salen* (Stockholm, 1754).

———: *Åminnelse-tal öfver slotts-byggmästaren och kongl. vetenskaps academiens ledamot, herr Claes Eliander, hållit den 27 februarii, 1756* (Stockholm, 1756).

Mukerji, Chandra: *Impossible engineering. Technology and territoriality on the Canal du Midi* (Princeton NJ, 2009).

———: *Territorial ambitions and the gardens of Versailles* (Cambridge, 1997).

Norberg, Axel: *Prästeståndets riksdagsprotokoll* vol. 6 [1723] (Stockholm, 1982).

———: *Prästeståndets riksdagsprotokoll* vol. 7 [1726–31] (Stockholm, 1975).

Nordbäck, Carola: *Samvetets röst. Om mötet mellan luthersk ortodoxi och konservativ pietism i 1720-talets Sverige* (Umeå, 2004).

Nordberg, Jöran Andersson: *Konung Carl den XII:tes historia* (Stockholm, 1740).

Nordencrantz, Anders: *Arcana oeconomiæ et commercii, eller Handelens och hushåldnings-wärkets hemligheter* (Stockholm, 1730).

Nordenflycht, Hedvig Charlotta: "Autobiographical letter to H:r A.A. von Stiernman, 1745-08-17", *Skrifter* (Stockholm, 1996).

Nordin, Jonas: *Frihetstidens monarki. Konungamakt och offentlighet i 1700-talets Sverige* (Stockholm, 2009).

Norrhem, Svante: *Uppkomlingarna. Kanslitjänstemännen i 1600-talets Sverige och Europa* (Umeå, 1993).

Nyberg, Klas: "'Jag existerar endast genom att äga kredit'. Tillit, kreditvärdighet och finansiella nätverk i 1700-talets och det tidiga 1800-talets Stockholm", in Mats Berglund (ed): *Sakta vi gå genom stan. Stadshistoriska studier tillägnade Lars Nilsson den 31/5 2005* (Stockholm, 2005), 184–211.

Oldenziel, Ruth: *Making technology masculine. Men, women, and modern machines in America, 1870–1945* (Amsterdam, 1999).

Olin, Martin: *Det karolinska porträttet. Ideologi, ikonografi, identitet* (Stockholm, 2000).

Ong, Walter J.: *Orality and literacy. The technologizing of the word* (London, 1991).

Ovid: *The XV bookes of P. Ouidius Naso, entytuled Metamorphosis, translated oute of Latin into English meeter, by Arthur Golding Gentleman, a worke very pleasaunt and delectable* (London, 1567).

Palmqvist, Fredrik: *Inledning til algebra. Första delen* (Stockholm, 1748).

———: *Underwisning i räkne-konsten* (Stockholm, 1750).

Persson, Fabian: *Servants of fortune. The Swedish court between 1598 and 1721* (Lund, 1999).

Persson, Mathias: *Det nära främmande. Svensk lärdom och politik i en tysk tidning 1753–1792* (Uppsala, 2009).

Peursen, C. A. Van: "E. W. von Tschirnhaus and the ars inveniendi", *Journal of the history of ideas* 54:3 (1993), 395–410.

Picon, Antoine: *French architects and engineers in the Age of Enlightenment* (New York, 1992).

Piirimäe, Pärtel: "The pen is a mighty sword. Johann Reinhold Patkul's polemical writings", *Die baltischen Länder und der Norden. Festschrift für Helmut Piirimäe zum 75. Geburtstag* (Tartu, 2005), 314–41.

Polhem, Christopher: "[Att Gud alzmächtig ähr hela naturens uphof…]", in Axel Liljencrantz (ed): *Christopher Polhems efterlämnade skrifter* vol. 3 (Uppsala & Stockholm, 1952), 304–15.

———: "Berättelse om Fahlu grufvas tillstånd", in Henrik Sandblad (ed): *Christopher Polhems efterlämnade skrifter* vol. 1 (Uppsala & Stockholm, 1947), 26–40.

———: "Lefvernesbeskrifning", in Bengt Löw (ed): *Christopher Polhems efterlämnade skrifter* vol. 4 (Uppsala & Stockholm, 1954 [1733]), 387–96.

———: "Mechanica practica eller fundamental byggarekonst", in Henrik Sandblad (ed): *Christopher Polhems efterlämnade skrifter* vol. 1 (Uppsala & Stockholm, 1947), 79–83.

———: "Samtahl emällan fröken Theoria och byggmästar Practicus om sitt förehafvande", in Henrik Sandblad (ed): *Christopher Polhems efterlämnade skrifter* vol. 1 (Uppsala & Stockholm, 1947), 277–307.

———: "Samtahl och discurs emellan theoria och praxis om mechaniska och physicalska saker huar igenom ungdomen, som der till har lust, kan lära något", in Axel Liljencrantz (ed): *Christopher Polhems efterlämnade skrifter* vol. 3 (Uppsala & Stockholm, 1952), 427–45.

———: *Samtal emellan en swär-moder och son-hustru, om allehanda hus-hålds förrättningar* (Stockholm, 1745).

———: "Tankar om mekaniken", *Kungliga vetenskapsakademins handlingar* 1 (1740), 185–98.

————: *Wishetens andra grundwahl til ungdoms prydnad mandoms nytto och ålderdoms nöje; lempadt för ungdomen efter theras tiltagande åhr, uti dagliga lexor fördelt* (Uppsala, 1716).

Raam, Peder Nilsson: *Then swenske åkermätningen, eller Ortuga deelo book, item een lijten tractat, om staaff och råå, och thes beskaffenheet* (Strengnäs, 1670).

Raeff, Marc: "The well-ordered police state and the development of modernity in seventeenth- and eighteenth-century Europe. An attempt at a comparative approach", *The American historical review* 80:5 (1975), 1221–43.

Ranft, Michael, and Johann Samuel Heinsius: *Die merkwürdige Lebensgeschichte derer vier berühmten schwedischen Feldmarschalle, Grafen Rehnschild, Steenbock, Meyerfeld und Dücker nebst dem angefügten merkwürdigen Leben und jämmerlichen Ende des bekannten Generals Johann Reinhold Patkuls. Zur Erleuterung vieler wichtigen Umstände der Geschichte Königs Caroli XII von Schweden, ans Licht gestället von einem Liebhaber der neuesten Historie* (Leipzig, 1753).

Rausing, Lisbet: "Daedalus Hyperboreus. Baltic natural history and mineralogy in the Enlightenment", in William Clark, Jan Golinski, and Simon Schaffer (eds): *The sciences in enlightened Europe* (Chicago, 1999), 389–422.

————: *Linnaeus. Nature and nation* (Cambridge MA, 1999).

————: "Underwriting the oeconomy. Linnaeus on nature and mind", in Margaret Schabas and Neil De Marchi (eds): *Oeconomies in the age of Newton* (Durham NC, 2004), 173–203.

Rimm, Stefan: *Vältalighet och mannafostran. Retorikutbildningen i svenska skolor och gymnasier 1724–1807* (Uppsala, 2011).

Rinman, Sven: *Åminnelse-tal öfver framledne bergmästaren … Axel Fred. Cronstedt, på kongl. academiens vägnar, hållet i stora riddarhus-salen den 6. martii, 1766* (Stockholm, 1766).

Roberts, Lissa: "Filling the space of possibilities. Eighteenth-century chemistry's transition from art to science", *Science in context* 6:02 (1993), 511–53.

————: "Full steam ahead. Entrepreneurial engineers as go-betweens during the late eighteenth century", in Simon Schaffer, Lissa Roberts, Kapil Raj, and James Delbourgo (eds): *The brokered world. Go-betweens and global intelligence, 1770–1820* (Sagamore Beach MA, 2009), 193–238.

————: "Geographies of steam. Mapping the entrepreneurial activities of steam engineers in France during the second half of the eighteenth century", *History and technology* 27:4 (2011), 417–39.

————: "Introduction. Workshops of the hand and mind", in Lissa Roberts, Simon Schaffer, and Peter Dear (eds): *The mindful hand. Inquiry and invention from the late Renaissance to early industrialisation* (Amsterdam & Bristol, 2007), 1–8.

Roberts, Lissa, Simon Schaffer, and Peter Dear (eds): *The mindful hand. Inquiry and invention from the late Renaissance to early industrialisation* (Amsterdam & Bristol, 2007).

Roberts, Michael: *The Age of Liberty. Sweden 1719–1772* (Cambridge, 1986).

Roche, Daniel: *France in the Enlightenment* (Cambridge MA, 1998).

————: *La France des Lumières* (Paris, 1993).

Rodhe, Staffan: *Matematikens utveckling i Sverige fram till 1731* (Uppsala, 2002).

Rohr, Julius Bernhard von: *Inledning til klokheten at lefwa, eller Underrättelse om, huru en menniskia, genom en förnuftig sitt lefwernes inrättning, kan bli timmeligen lycksalig* (Stockholm, 1728).

Roman, Christian Gustafsson: *Wälmente tanckar om barna upfostring. I anseende till deras studier och lefnad* (Västerås, 1743).

Roper, Lyndal: *Oedipus and the Devil. Witchcraft, sexuality and religion in early modern Europe* (London, 1994).

Roper, Michael, and John Tosh: "Introduction", in Michael Roper and John Tosh (eds.) *Manful assertions. Masculinities in Britain since 1800* (London, 1991), 1–24.

———: *Manful assertions. Masculinities in Britain since 1800* (London, 1991).

Roux, Sophie, and Walter Roy Laird: "Introduction", in Walter Roy Laird and Sophie Roux (eds): *Mechanics and natural philosophy before the scientific revolution* (Dordrecht, 2007), 1–11.

Ruder, Gustav: *Anledning til snille-walet, eller Ungdomens snille-gåfwors och naturliga böjelsers bepröfwande, wal och anförande, til the wetenskaper, konster, ämbeten och syslor, med hwilka hwar och en kan wäl och beqwämligen tjena Gudi och fädernes landet* (Stockholm, 1737).

Runefelt, Leif: *Dygden som välståndets grund. Dygd, nytta och egennytta i frihetstidens ekonomiska tänkande* (Stockholm, 2005).

———: *Hushållningens dygder. Affektlära, hushållningslära och ekonomiskt tänkande under svensk stormaktstid* (Stockholm, 2001).

Rydelius, Andreas: *Nödige förnufftz öfningar för all slags studerande ungdom, som wil hafwa sunda tankar, och fälla ett billigt omdöme om de högste och wichtigste ting i wärlden, hwilket i naturlig måtta bör wara deras ypperesta ändamål, ehwad för stånds wahl de hälst willja giöra* 5 vols. (Linköping, 1718–22).

Rydén, Göran: "Skill and technical change in the Swedish iron industry, 1750–1860", *Technology and culture* 39:3 (1998), 383–407.

———: "The Enlightenment in practice. Swedish travellers and knowledge about the metal trades", *Sjuttonhundratal* (2013), 63–86.

Salvius, Lars: *Tanckar öfwer den swenska oeconomien igenom samtal yttrade [...] emellan fru Swea, fru Oeconomia, Herr Mentor, Herr Flit, Herr Sparsam, och Frans.* (Stockholm, 1738).

Savin, Kristiina: *Fortunas klädnader. Lycka, olycka och risk i det tidigmoderna Sverige* (Lund, 2011).

Sawday, Jonathan: *Engines of the imagination. Renaissance culture and the rise of the machine* (London, 2007).

Schabas, Margaret, and Neil De Marchi: "Introduction", in Margaret Schabas and Neil De Marchi (eds): *Oeconomies in the age of Newton* (Durham NC, 2004), 1–13.

———: *Oeconomies in the age of Newton* (Durham NC, 2004).

Schaffer, Simon: "Enlightened Automata", in William Clark, Jan Golinski, and Simon Schaffer (eds): *The sciences in enlightened Europe* (Chicago, 1999), 126–65.

———: "Experimenters' techniques, dyers' hands, and the electric planetarium", *Isis* 88:3 (1997), 456–83.

———: "Swedenborg's lunars", *Annals of science* 71:1 (2014), 2–26.

Schatzberg, Eric: "Technik comes to America. Changing meanings of technology before 1930", *Technology and culture* 47:3 (2006), 486–512.

Schauman, Georg: *Studier i frihetstidens nationalekonomiska litteratur. Idéer och strömningar 1718–1740* (Helsinki, 1910).

Schiefsky, Mark J.: "Theory and practice in Heron's Mechanics", in Walter Roy Laird and Sophie Roux (eds): *Mechanics and natural philosophy before the scientific revolution* (Dordrecht, 2007), 15–50.

Schmitt, Frederick F.: *Socializing epistemology. The social dimensions of knowledge* (Lanham MD, 1994).

Schmoll, Barthold Otto: *Kort anledning till geometrien, hwar effter officerarne wid Hans Kongl. May:tz infanterie här i Swerige till mathesin och fortification* (Stockholm, 1692).

Schück, Henrik: *Ansatser till ett universitet i Stockholm före 1800-talet* (Stockholm, 1942).

Schwab, Dieter: "Familie", *Geschichtliche Grundbegriffe. Historishes Lexicon zur politischen-sozialen Sprache in Deutschland* (Stuttgart, 1975).

Scott, Joan W.: "Gender. A useful category of historical analysis", *The American historical review* 91:5 (1986), 1053–75.

Semler, Christoph: *Neueröffnete mathematische und mechanische Real-Schule* (Halle, 1709).

Sennefelt, Karin: *Politikens hjärta. Medborgarskap, manlighet och plats i frihetstidens Stockholm* (Stockholm, 2011).

Sewell, William H.: "Visions of labor. Illustrations of the mehcanical arts before, in, and after Diderot's Encyclopedie", in Steven L. Kaplan and Cynthia J. Koepp (eds): *Work in France. Representations, meaning, organization, and practice* (Ithaca, 1986), 258–86.

Shapin, Steven: *A social history of truth. Civility and science in seventeenth-century England* (Chicago, 1994).

———: "The invisible technician", *American scientist* 77:6 (1989), 554–63.

———: *The scientific revolution* (Chicago, 1996).

Shapin, Steven, and Simon Schaffer: *Leviathan and the air-pump. Hobbes, Boyle, and the experimental life* (Princeton NJ, 1985).

Shepard, Alexandra: *Meanings of manhood in early modern England* (Oxford, 2003).

Sibum, H. Otto: "Experiencing experiment. Gestural knowledge and scientific change in early 19th-century Victorian culture", *Thought & culture* 10 (2011), 38–55.

———: "Experimentalists in the Republic of Letters", *Science in Context* 16:1–2 (2003), 89–120.

———: "Les gestes de la mesure, Joule, les pratiques de la brasserie et la science", *Annales HSS* July–October, 1998.

———: "Reworking the mechanical value of heat. Instruments of precision and gestures of accuracy in early Victorian England", *Studies in history and philosophy of science. Part A* 26:1 (1995), 73–106.

Siljestrand, Karl K:son: *Karl XII såsom filosof* (Linköping, 1891).

Sjöstrand, Wilhelm: *Pedagogikens historia* 4 vols. (Lund, 1954–66).

Smith, Jay M.: *The culture of merit. Nobility, royal service, and the making of absolute monarchy in France, 1600–1789* (Ann Arbor, 1996).

Smith, Pamela H.: *The body of the artisan. Art and experience in the scientific revolution* (Chicago, 2004).

Snickare, Mårten: *Enväldets riter. Kungliga fester och ceremonier i gestaltning av Nicodemus Tessin den yngre* (Stockholm, 1999).

———: "Shaping the ritual space. Nicodemus Tessin the younger and Swedish royal ceremonies", in Allan Ellenius (ed): *Baroque dreams. Art and vision in Sweden in the era of greatness* (Uppsala, 2003), 124–50.

Spöring, Herman Diedrich: *Johannis Moræi ... äre-minne, i auditorio illustri på riddarehuset efter kongl. wetenskaps academiens befallning uprättadt* (Stockholm, 1743).

Stadin, Kekke: "Att vara god eller att göra sin plikt? Dygd och genus i 1600-talets Sverige", in Janne Backlund (ed): *Historiska etyder* (Uppsala, 1997), 223–35.

———: *Stånd och genus i stormaktstidens Sverige* (Lund, 2004).

———: "The masculine image of a great power. Representations of Swedish imperial power c. 1630–1690", *Scandinavian Journal of History* 30:1 (2005), 61–82.

Staf, Nils: *Borgarståndets riksdagsprotokoll från frihetstidens början* vol. 2 [1723] (Stockholm, 1951).

———: *Borgarståndets riksdagsprotokoll från frihetstidens början* vol. 4 [1731] (Stockholm, 1958).

Stewart, Larry R.: *Rise of public science. Rhetoric, technology and natural philosophy in Newtonian Britain, 1660–1750* (Cambridge, 1992).

Strömer, Mårten: *Åminnelse-tal öfver kongl. maj:ts troman, stats-secreteraren ... Samuel Klingenstjerna, på k. vetensk. academiens vägnar hållit, den 27 jul. 1768* (Stockholm, 1768).

Rudolf L. Tafel (ed): *Documents concerning the life and character of Emanuel Swedenborg* vol. 1 (London, 1875).

Thanner, Lennart: *Revolutionen i Sverige efter Karl XII:s död. Den inrepolitiska maktkampen under tidigare delen av Ulrika Eleonora d.y:s regering* (Uppsala, 1953).

Theofrastos, and Abraham Magni Sahlstedt: *Caracterer, eller Sede-bilder af människor* (Stockholm, 1754).

Thomasius, Christian: *Gründliche iedoch bescheidene Deduction der Unschuld Hn. Joh. Reinhold von Patkul* (Leipzig, 1701).

Tilas, Daniel: *Curriculum vitæ I–II, 1712–1757 samt fragment av dagbok september–oktober 1767*, Holger Wichman (ed) (Stockholm, 1966 [1712–57]).

Tillaeus, Petrus: "General charta öfwer Stockholm med malmarne åhr 1733" (Stockholm, 1733).

Tjeder, David: "Konsten att blifva herre öfver hvarje lidelse. Den ständigt hotade manligheten", in Anne Marie Berggren (ed): *Manligt och omanligt i ett historiskt perspektiv* (Stockholm, 1999), 177–96.

Tosh, John: "What should historians do with masculinity? Reflections on nineteenth-century Britain", *History workshop* 38 (1994), 179–202.

Tribe, Keith: "Cameralism and the science of government", *The journal of modern history* 56:2 (1984), 263–84.

———: *Governing economy. The reformation of German economic discourse, 1750–1840* (Cambridge, 1988).

Tschirnhausen, Ehrenfried Walther von: *Gründliche Anleitung zu nützlichen Wissenschafften absonderlich zu der Mathesi und Physica. Wie sie anitzo von den Gelehrtesten abgehandelt werden* (Frankfurt & Leipzig, 1700).

Uhr, Isaac Johan: *En brukspatrons egenskaper* (Uppsala, 1750).

Ulff, Samuel Crispin: *En liten doch utförlig grundritning och handledning, til de metall- och linne-manufacturier, som nu inrättas i Hälsinge-land, at drifwas på åtskilligt nytt arbetz sätt, särdeles af landetz egne producter; hwilka under twenne societeter komma at sortera, såsom deras ägare och förläggare: blifwandes här ock något förestält om manufacturerne i gemen* (Stockholm, 1729).

Vérin, Hélène: *La gloire des ingénieurs. L'intelligence technique du XVIe au XVIIIe siècle* (Paris, 1993).

Voltaire, François Marie Arouet de: *Histoire de Charles XII, roi de Suède, divisée en huit livres, avec l'histoire de l'empire de Russie sous Pierre-le-Grand, en deux parties divisées par chapitres* (Geneva, 1768).

Wahlström, Anders: *Ethica mathematica. Heller mathematisk sedelära* (Västerås, 1742).

Wakefield, Andre: "Books, bureaus, and the historiography of cameralism", *European journal of law and economics* 19:3 (2005), 311–20.

———: "Butterfield's nightmare. The history of science as Disney history", *History and technology* 30:3 (2014), 232–51.

———: "Leibniz and the wind machines", *Osiris* 25:1 (2010), 171–88.

———: *The disordered police state. German cameralism as science and practice* (Chicago, 2009).

Wallerius, Göran: *Tal emellan mathesin och physiquen om deras verkan och nytta uti bärgs-väsendet, förestäldt i kongl. svenska vetenskaps academien af Göran Vallerius vid præsidii afläggande den 21 jan. 1744* (Stockholm, 1747).

Wargentin, Pehr Wilhelm: *Åminnelse-tal öfver kongl. observatoren och kongl. vetenskaps academiens ledamot, herr Olof Hiorter, efter kongl. vetenskaps academiens befallning, hållit i stora riddarhus-salen d. 18 april, 1751* (Stockholm, 1751).

———: *Åminnelse-tal öfver kongl. vetensk. academiens framledne ledamot, directeuren och mathematiska instrument-makaren, herr Daniel Ekström, hållit för kongl. vetenskaps academien, den 14 junii år 1758* (Stockholm, 1758).

Warwick, Andrew: "Exercising the student body. Mathematics and athleticism in Victorian Cambridge", in Christopher Lawrence and Steven Shapin (eds): *Science incarnate. Historical embodiments of natural knowledge* (Chicago, 1998), 288–323.

———: *Masters of theory. Cambridge and the rise of mathematical physics* (Chicago, 2003).

Waters, Malcolm: "Patriarchy and viriarchy. An exploration and reconstruction of concepts of masculine domination", *Sociology* 23:2 (1989), 193–211.

Weidler, Johann Friedrich: *Institutiones mathematicae decem et sex pvrae mixtaeque matheseos disciplinas complexae* (Wittenberg, 1718).

Weidler, Johann Friedrich and Johan Mört (tran.): *En klar och tydelig genstig eller anledning til geometrien och trigonometrien* (Stockholm, 1727).

Wermlund, Sven: *Sensus internus och sensus intimus. Studier i Andreas Rydelius filosofi* (Stockholm, 1944).

Westfall, Richard S.: "The rise of science and the decline of orthodox Christianity. A study of Kepler, Descartes, and Newton", in David C. Lindberg and Ronald L. Numbers (eds): *God and nature. Historical essays on the encounter between Christianity and science* (Berkeley, 1986).

Whitmer, Kelly J.: "Eclecticism and the technologies of discernment in pietist pedagogy", *Journal of the history of ideas* 70:4 (2009), 545–67.

———: *The Halle Orphanage as scientific community. Observation, eclecticism and pietism in the early enlightenment* (Chicago, 2015).

Wiberg, Albert: *Christopher Polhem som slöjdpedagog* (Gothenburg, 1938).

———: *Till skolslöjdens förhistoria. Några utvecklingslinjer i svensk arbetspedagogik intill 1877. Tiden intill 1861* vol. 1 (Stockholm, 1939).

Widmalm, Sven: "Instituting science in Sweden", in Roy Porter and Mikuláš Teich (eds): *The scientific revolution in national context* (Cambridge, 1992), 240–62.

———: *Mellan kartan och verkligheten. Geodesi och kartläggning, 1695–1860* (Uppsala, 1990).

Willmoth, Frances, and Rob Iliffe: "Astronomy and the Domestic Sphere. Margaret Flamsteed and Caroline Herschel as Assistant-Astronomers", in Lynette Hunter and Sarah Hutton (eds): *Women, science and medicine 1500–1700. Mothers and sisters of the Royal Society* (Stroud, 1997), 235–65.

Winton, Patrik: *Frihetstidens politiska praktik. Nätverk och offentlighet 1746–1766* (Uppsala, 2006).

Wise, M. Norton: "Mediating machines", *Science in context* 2:1 (1988), 77–113.

Wolff, Christian von and Johan Mört (tran.): *Christian Wolffs Grundeliga underrättelse om rätta orsaken til sädens vnderbara förökelse, vti hwilken tillika förklaras träns och plantors wäxt i gemen; såsom första profwet, af vndersökandet om plantors wäxt, på tyska först vtgifwen; men sedermera alla hushållare i gemen, och i synnerhet Ultuna laborator. til tienst, på swenska öfwersatt, och med hans kongl. maj:ts allernådigste privilegio vplagd* (Stockholm, 1728).

Wrangel, Ewert: *Frihetstidens odlingshistoria ur litteraturens häfder 1718–1733* (Lund, 1895).

Wright, Thomas: *William Harvey. A life in circulation* (Oxford, 2012).

Zanger, Abby E.: *Scenes from the marriage of Louis XIV. Nuptial fictions and the making of absolutist power* (Stanford CA, 1997).

Zetherström, Johan Niclas: *Rikets nytta af välbelönte ämbetsmän, tänkande informatorer och utvalde studerande* (Stockholm, 1765).

Zetterberg, J. Peter: "The mistaking of 'the mathematicks' for magic in Tudor and Stuart England", *The sixteenth century journal* 11:1 (1980), 83–97.

Öhrberg, Ann: *Samtalets retorik. Belevade kulturer och offentlig kommunikation i svenskt 1700-tal* (Höör, 2014).

Index of names

Institutionen för idé- och lärdomshistoria
Uppsala universitet
Uppsala Studies in History of Ideas (Nr 1-32: Skrifter)

Editors: Sven Widmalm och H. Otto Sibum

1. Per-Gunnar Ottosson, Scholastic Medicine and Philosophy: A Study of Commentaries of Galen's Tegni (ca. 1300–1450) (1982)
2. Franz Luttenberger, Neuroser och neuroterapi ca. 1880–1914 (1982)
3. Roy Porter, The History of Medicine: Past, Present, and Future (1983)
4. Tore Frängsmyr, Liten handbok för avhandlings- och uppsatsskrivare (1983)
5. Anders Burius, Ömhet om friheten: Studier i frihetstidens censurpolitik (1984)
6. Per-Gunnar Ottosson, Synen på pesten: Exempel och problem från svensk stormaktstid (1984)
7. Hippokrates i urval, med inledning av Eyvind Bastholm (1984)
8. Franz Luttenberger, Freud i Sverige: Psykoanalysens mottagande i svensk medicin och idédebatt (1988)
9. Tore Frängsmyr, Science or History: Georg Sarton and the Positivist Tradition in the History of Science (1989). Nytryck ur Lychnos 1973/74
10. Sven Widmalm, Mellan kartan och verkligheten: Geodesi och kartläggning, 1695–1860 (1990)
11. Lars Sellberg, Av kärlek till Fosterland och Folk: Gabriel Djurklou och dialektforskningen (1993)
12. Urban Wråkberg, Vetenskapens vikingatåg: Perspektiv på svensk polarforskning 1860–1930 (1995). Ny upplaga 1999.
13. Thomas H. Brobjer, Nietzsche's Ethics of Character: A Study of Nietzsche's Ethics and its Place in the History of Moral Thinking (1995)
14. Per Dahl, Svensk ingenjörskonst under stormaktstiden: Olof Rudbecks tekniska undervisning och praktiska verksamhet (1995)
15. Suzanne Gieser, Den innersta kärnan: Djuppsykologi och kvantfysik, Wolfgang Paulis dialog med C.G. Jung (1995)
16. Mikael Hörnqvist, Machiavelli and the Romans (1996)
17. Tore Frängsmyr, ed., Nordström och hans skola: Bakgrund – nuläge – utveckling (1997)
18. Tore Frängsmyr, Den gudomliga ekonomin: Religion och hushållning i 1700-talets Sverige (1997)
19. Gunnar Matti, Det intuitiva livet: Hans Larssons vision om enhet i en splittrad tid (1999)
20. Olov Amelin, Medaljens baksida: Instrumentmakaren Daniel Ekström och hans efterföljare i 1700-talets Sverige (1999)

21. Gunnar Eriksson & Karin Johannisson, eds., Den akademiska gemenskapen: Universitetets idé och identitet, ett symposium till Tore Frängsmyrs 60-årsdag (1999)
22. Karl Grandin, Ett slags modernism i vetenskapen: Teoretisk fysik i Sverige under 1920-talet (1999)
23. Martin Bergström, Anders Ekström & Frans Lundgren, Publika kulturer: Att tilltala allmänheten, 1700-1900. En inledning (2000)
24. Karin Johannisson, Naturvetenskap på reträtt: En diskussion om naturvetenskapens status under svenskt 1700-tal (2001). Nytryck ur Lychnos 1979/80
25. Gunnar Eriksson, Olof Rudbeck d.ä. (2001). Nytryck ur Lychnos 1984
26. Adrian Thomasson, En filosof i exil: Ernst Cassirer i Sverige, 1935–41 (2001)
27. Urban Josefsson, Det romantiska tidehvarfvet: De svenska romantikernas medeltidsuppfattning (2002)
28. Olof Ljungström, Oscariansk antropologi: Exotism, förhistoria och rasforskning som vetenskaplig människosyn (2002)
29. Oskar Pettersson, Politisk vetenskap och vetenskaplig politik: Studier i svensk statsvetenskap kring 1900 (2003)
30. Hjalmar Fors, Mutual Favours: The Social and Scientific Practice of Eighteenth-Century Swedish Chemistry (2003)
31. Ulrika Nilsson, Kampen om Kvinnan: Professionaliseringsprocesser och konstruktioner av kön i svensk gynekologi 1860–1925 (2003)
32. Peter Josephson, Den akademiska frihetens gränser: Max Weber, Humboldtmodellen och den värdefria vetenskapen (2005)
33. Daniel Lövheim, Att inteckna framtiden: Läroplansdebatter gällande naturvetenskap, matematik och teknik i svenska allmänna läroverk 1900–1965 (2006)
34. Shamal Kaveh, Det villkorade tillståndet: Centralförbundet för Socialt Arbete och liberal politisk rationalitet 1901–1921 (2006)
35. Tobias Dahlkvist, Nietzsche and the Philosophy of Pessimism: A Study of Nietzsche's Relation to the Pessimistic Tradition: Schopenhauer, Hartmann, Leopardi (2007)
36. Tony Gustafsson, Läkaren, döden och brottet: Studier i den svenska rättsmedicinens etablering (2007)
37. Emma Shirran, Samhälle, vetenskap och obstetrik: Elis Essen-Möller och kvinnokliniken i Lund (2007)
38. Torbjörn Gustafsson Chorell, Fascination (2008)
39. Annika Berg, Den gränslösa hälsan: Signe och Axel Höjer, folkhälsan och expertisen (2009)
40. Mathias Persson, Det nära främmande: Svensk lärdom och politik i en tysk tidning, 1753–1792 (2009)
41. Louise Nilsson, Färger, former, ljus. Svensk reklam och reklampsykologi, 1900–1930 (2010)

42. Jenny Ingemarsdotter, Ramism, Rhetoric and Reform. An Intellectual Biography of Johan Skytte (1577–1645) (2011).
43. Maria Björk, Problemet utan namn? Neuroser, stress och kön i Sverige från 1950 till 1980. (The Problem that had no Name? Neurosis, Stress and Gender in Sweden 1950-1980) (2011)
44. David Thorsén, Den svenska aidsepidemin. Ankomst, bemötande, innebörd. (2013)
45. My Klockar Linder, Kulturpolitik. Formeringen av en modern kategori (2014)
46. Jacob Orrje, Mechanicus. Performing an Early Modern Persona (2015)

Beställningar gällande nr. 1–32 skall ställas till:
Inst. för idé- och lärdomshistoria
Uppsala universitet
Box 629
751 26 Uppsala

Beställningar gällande fr.o.m. nr. 33 skall ställas till:
Uppsala universitetsbibliotek
Box 510
751 20 Uppsala

ISSN 1653-5197